Designing for Life

Pertti Saariluoma • José J. Cañas • Jaana Leikas

Designing for Life

A Human Perspective on Technology Development

palgrave
macmillan

Pertti Saariluoma
Cognitive Science
University of Jyväskylä, Finland

José J. Cañas
Faculty of Psychology
University of Granada
Granada, Spain

Jaana Leikas
VTT Technical Research Centre
of Finland Ltd
Tampere, Finland

ISBN 978-1-137-53046-2   ISBN 978-1-137-53047-9   (eBook)
DOI 10.1057/978-1-137-53047-9

Library of Congress Control Number: 2016942071

Cover illustration: © Garry Gay / Alamy Stock Photo

Printed on acid-free paper

This Palgrave Macmillan imprint is published by Springer Nature
The registered company is Macmillan Publishers Ltd. London

Preface

Technology is only valuable to the extent that it can enhance the quality of life. When solving complex engineering problems, it is easy to forget the basic reason why technologies are designed and developed. It is also easy to forget what actually turns a technical idea into an innovation. Although designers create technologies, only users can, in the end, decide what becomes an innovation.

This book was written to expound on the true value of taking human life as the starting and focus point for technology design. Given the narrow interpretation of the concept of 'human', few definite design practices include the perspective of human life. As information and communication technologies have proliferated vastly beyond the scope of tools and become a ubiquitous part of people's daily lives, the need for proper design approaches has become obvious. The traditional focus on human-centred design approaches is increasingly considered insufficient for understanding human–technology interaction (HTI), as it arises from an interest in technology rather than an interest in humans. It has been recognized that completely automated technologies do not exist, and designers have accepted that there will always be a human interacting with technologies. However, 'human' in this sense has meant that a human being, the user, has been the focus only as a test object for the design and launch of a desired technology. Accordingly, the most-emphasized aspect of HTI has been whether people find it easy to use certain products and systems, as difficult-to-use technology is easily rejected. From this perspective, the field of technology and innovation design can be seen more as an engineering discipline in which, for example, psychological theories of interaction are employed first and foremost as instruments for developing 'usable' technology.

There have been attempts to change this practice to make it more human centred by introducing the term 'human-driven design', where the starting point in a design should be the human being and his or her needs, goals, and desires. The focus has been transferred from technology and 'quality assurance' to a more human approach, and the significance and emphasis of the human sciences in design has increased. Accordingly, the practical implications of this field (and the resources channelled to it) have been substantial.

Yet interpretations of the concept of 'human' have been narrow, and perspectives on how to understand people's needs have been limited—and have varied depending on the purpose and the interpreter. They have provided no concepts that would make it possible to study the real needs of people that could be fulfilled with desirable, sustainable, and ethically accepted technology.

Why has it been so difficult to design the human dimension of technology? We argue that this is because the most profound factor of HTI has been left out of the discussion: people's everyday lives. Understanding daily life makes it possible to create a practical methodology that can be used effectively to design the human dimension of technology, and to convince business developers of the true value of treating the human perspective as a cornerstone of the design. Quite simply, this means that HTI design should be able to perceive, analyse, and design technology through knowledge of (and for) people's everyday lives.

Designing technology to improve the quality of human life requires a multidisciplinary design approach. On one hand, multidisciplinary teams can give designers with a technical background the opportunity to better acquaint themselves with human research by working with human researchers. On the other hand, human researchers should be more aware of the various roles they can play in the process of designing and developing new technological solutions for people. Human researchers can be provided with concepts, facts, methods, and theories that are useful in many aspects of design.

This book is written for a wide group of professionals who may look at design through very different types of conceptual lenses. For engineering designers, it may strengthen their understanding of human research available for developing HTI design. For human researchers from psychologists to physiatrists, and from sociologists, anthropologists, to artists and historians, this book should give an idea of how to apply and strengthen their expertise in the discourses of developing human life by means of technical design and development. As said, design thinking is a multidisciplinary activity that requires different roles to be filled with different kinds of experts.

After recognizing the multidisciplinary nature of HTI design, the question is how knowledge about humans can be incorporated into design processes efficiently. This book introduces a new way of organizing creative thinking about design, and discusses scientific knowledge about the human mind as an essential element of design discourses. It reviews different perspectives of HTI and contributes ideas of life-based design. It systematizes traditional design discourses by showing that HTI design thinking must always meet four fundamental design questions:

1. Functionalities and technical user interface design;
2. Fluency and ease of use;
3. Elements of experience and liking; and
4. Position of technology in human life.

As these fundamental questions are always present when designing the human dimension of technology, it is useful to understand the logic behind them. The questions define the basic tasks in HTI design: the decision of the functionalities of the artefact, understanding how to best use them, understanding the overall experience when using them, and finally—and most importantly—the technology's role in human life. The latter is always present in design, either consciously or tacitly.

Finally, the book shows that it is possible to build ontological systems around these fundamental questions to rationally organize the management of HTI design processes and to integrate the questions into a unified research and design process.

We start by discussing the mutually inclusive roles of the natural sciences and human research in design, and the foundations of (and relationship between) scientific and design thinking. Chapter 2 introduces the main paradigms of HTI design using different approaches, perspectives, concepts, and methodologies. It reveals that HTI research and design is dominated by the logic of questions, and discusses the basic questions that provide a fundamental structure to the field. In Chap. 3 we discuss the logic of user interface design and present the overall principles and goals for the user interface design of any technical artefact. Chapter 4 discusses the importance of understanding the cognitive elements and psychological and mental preconditions for using technology—that is, the psychology of fluent use. Chapter 5 examines aspects of the dynamic mind—in particular human emotions, motives, and personality—that help address people's preferences and how they feel when they use technologies. Chapter 6 introduces the reader to the conceptual model of life-based design, which is based on segregating unified systems of action called forms of life. Investigating the structure of actions and related facts relevant to particular forms of life, in addition to the values that people adhere to, constitutes the core tool of this design approach. The knowledge produced comprises a template for human requirements, which serves as a basis for design ideas and technological solutions. Chapter 7 discusses product design as a constructive process, which requires setting up and testing hypotheses and developing logical chains of arguments to decide which assumptions are facts and which are not. It involves creating new perspectives for searching out truths by designing more accurate concepts, which allow new types of questions to be asked concerning the state of affairs. Finally, the chapter introduces generic design ontology for HTI design, and concludes by exploring the possibility of truly fitting design solutions into people's everyday lives and advancing the quality of life by ethical design.

We would like to thank Jyväskylä and Granada Universities as well as VTT Technical Research Centre of Finland for being stimulating research and innovation environments. We also wish to thank the Design United and Department of Industrial Design at Eindhoven Technical University for offering an inspiring environment for the first author during a visit there. Without this opportunity, this book would still be unfinished. The Finnish authors wish to thank the National Technology Agency (Tekes) for the Theseus I-II and ITEA-2 for Easy Interactions projects, which facilitated the start of this work. Tekes also supported the BeWell and Life-Based Design projects, which have been important for developing the basic conceptions of the book. The work of the Spanish author has been supported by projects SEJ2007-63850/PSIC and PSI2012-39246 from the Spanish Ministry of Education and Science, and project ProyectoMIO! from the Spanish Ministry of Industry. We thank PalgraveMacMillan Team, Nicola Jones, Eleanor Christie, Sharla Plant, Greg Willliams, Ramaswamy Parvathy, and Angeline Amrita Stanley for their invaluable help. We also thank Kelley Friel for correcting the language. Finally, we thank Noora Hirvonen for the illustrations.

We dedicate this book to the light nights of Finnish summers and the gentle Andalusian evenings, when the true Finnish–Spanish cooperation took place in the heat of a Finnish sauna and in the blue hills of Granada.

Pertti Saariluoma

José J. Cañas

Jaana Leikas

Virrat, Finland

June 2015

Contents

List of Figures

List of Tables

1

Technology in Life

Technological advancements have changed human life throughout history as technical inventions have emancipated people from many mundane, necessary tasks. The development of technical artefacts has long relied on the natural sciences and engineering. However, recent technical advancements—such as ubiquitous and massively multifunctional technologies as well as the emergence of social media—have made it necessary to approach design from a multidisciplinary perspective and to ground design thinking more on the understanding of human mind and human life. As the natural sciences and human research are in many respects different practices, it is time to discuss their mutually inclusive roles in design and to rethink the foundations of (and relationship between) scientific and design thinking.

Information and communication technology (ICT) services have become an increasingly central form of social activity, which is radically changing the foundations of technology design thinking. Human–technology interaction (HTI) design no longer involves merely developing new technical devices or artefacts for people to use. It is shifting from creating tools to shaping what they can *achieve*. For example, hundreds of thousands of apps and an unlimited number of web pages have been developed in recent years without making any essential changes to the

© The Editor(s) (if applicable) and The Author(s) 2016   **1**
P. Saariluoma et al., *Designing for Life*,
DOI 10.1057/978-1-137-53047-9_1

technical artefacts on which the apps are used. Moreover, technical solutions are becoming more complex, and automatic processes have replaced many operators. Though creating new artefacts never ceases, it is evident that the ever-evolving human relationship with technology is becoming more and more important.

The term 'technical artefact' or 'artefact' refers to any human-made object, natural process, or even a modified natural phenomenon that is used to improve human performance, satisfy some human need, or improve the quality of life (Simon 1969). An artefact can also be a technical machine or device that has a human function (Houkes et al. 2011). It can be a computer program, machine, device, instrument, chemical product, or tool. It can also be something that gives subjective enjoyment to people—a film, a TV programme, or a cigarette. It can even be a natural phenomenon used by people, like fire or a stone tool. The main requirement is it helps people pursue their action goals. The term 'technology' typically refers to a combination of technical artefacts and their human uses—that is, what people do with artefacts and how they are organized around them (Davis et al. 2014; Houkes et al. 2011; Karwowski 2006; Meier-Oeser 1998; Orlikowski 1991, 2000).

People, and the various roles they play when interacting with technology—such as users, consumers, and operators—constitute the human dimension of technology. HTI in its broadest sense covers all forms and aspects of interaction between people and technical artefacts; it includes the roles of designer, business manager, object of action, constructor, and builder.

Although issues related to computational devices dominate today's discussion on HTI, traditional mechanical artefacts and their uses are also relevant for this discussion, since most technical artefacts today have both computational and mechanical dimensions. The nature of scientific knowledge in design thinking is changing, and designers and developers of modern technologies are required to acquire new types of skills. For centuries, natural science and mathematics have provided the basic concepts and theories for technology design thinking and innovation, but they are no longer sufficient in HTI design. The interaction between humans and new types of technologies must be holistically understood in all its complexity.

The human dimension is an inevitable part of technology design thinking, since technological solutions have direct or indirect links to everyday life. For example, modern paper machines can dry paper incredibly quickly, thanks to the invention of the so-called extended nip, a flexible mantle that sped up the process considerably and improved the quality of paper (Saariluoma et al. 2006). Therefore, this invention increased the productivity of papermaking factories by tens of percentages, which improved the quality of life for the users of the paper (e.g., newspaper readers) and increased investors' return of investment (ROI).

In engineering, design thinking is conceived as organizing the laws of nature in a meaningful manner (Pahl et al. 2007). Also, understanding the human mind and human life is necessary for people working with technology design. However, the traditional (largely common sense and user-need based) conceptualization of the human mind and actions is no longer sufficient. The conceptual structure of design thinking has to be expanded to also consider other concepts of human research (Davis et al. 2014; Houkes et al. 2011; Karwowski 2006; Meier-Oeser 1998; Orlikowski 1991, 2000).

The most important design solutions are based on scientific knowledge developed in the natural sciences such as mathematics, physics, chemistry, and materials science. Since they are able to apply the laws of nature and to use simulations to study different alternatives, engineers can confidently predict that their design solutions will work in practice. Human research, by contrast, is not applied as effectively as the laws of natural science in technology design. Thus, while a design might perform well technically, users may not be able to (or may not want to) use the artefact, or simply have no practical use for it.

As the human dimension of technology is becoming more important in practical design, development, and innovation, it is necessary to turn one's attention to human research, that is, the relevant areas of biology, psychology, and socio-cultural research. Yet natural scientific and human perspectives on thinking are conceptually different in many respects. While human research is dominated by causal explanation, the natural sciences rely on such philosophical and metascientific notions as intention or understanding (Radnitsky 1968; von Wright 1971).

There has been relatively little communication between the academic human–computer interaction (HCI) community and industry in the area of HTI design (Carroll 1997), and researchers in both fields have approached interaction issues very differently. Consequently, the degree of applying human research is not as high as it could be in modern technology design.

Engineering designers and programmers are experts in the natural sciences and mathematics, while human researchers are qualified in human biology, psychology, and socio-cultural research. Human researchers have used different concepts and seen different problems than technical designers, and industry professionals have felt that human researchers do not comprehend the real goal of design thinking, which is working with a product rather than coming up with a good theory.

Indeed, people are goal oriented rather than abstractly theoretical when it comes to using technology (Card et al. 1983). People, moreover, are different. Though designers have expected them to behave like machine parts with specific functions, from registering odd things in the environment to typing in texts, they have dimensions that should also be factored in. People err, they change their moods and goals, they feel good or bad, and they have their life histories, social networks, and personality. But most of all, they have their own individual lives. Since all technologies are designed to be used in life and are justified (or not) by their significance in human life, technology designers cannot ignore people and their lives.

Incorporating human knowledge into technology design thinking thus uncovers underlying conceptual structures and tacitly or explicitly guides design thinking. Numerous paradigms have been developed to incorporate the human dimension of technology. Common examples include usability testing, user experience evaluation, and usability goal setting (Dumas and Redish 1999; Hassenzahl and Tractinsky 2006; Helander and Khalid 2006; Nagamashi 2011; Nielsen 1993; Wixon and Wilson 1997). Designers have also constructed sketches, mock-ups, simulation models, and prototypes, which they more or less systematically test with users to find out whether they make errors or take more time than necessary to perform certain tasks. However, the kind of approach that does not fully take human research into account when creating design goals and requirements can be seen as tacitly conservative, because it prioritizes technical advancement over the human mind and people's lives.

A number of different kinds of tools have been developed to support design practices, such as graphical user interface components, style guides, standards, practices, and databanks (Cooper et al. 2007; Collison et al. 2009; Goodwin 2011; ISO 1998a, b; Kim et al. 2011). These tools can be effective when they are used to solve specific and restricted design details, but when misapplied or used superficially, they overlook deeper systematizations or understanding of how users' minds work (Norman 2002).

From a higher conceptual level more focused on the design of HTI, the critical step is to create and describe the conceptual foundations of new, human research-based design thinking. This would require defining the main interaction design and research concepts, problems, and organization. It would also presuppose describing the design processes, the nature of the designers' thinking, and the logic that unifies the human research and natural scientific design thinking in HTI design practices. This new way of thinking should lead to a transparent and explicit understanding of how human research can support designers' thinking in HTI design and innovation. From this perspective, the goal of design is not to create new technical artefacts but to change how people live and to improve the quality of human life. The whole history of technology justifies this perspective, and therefore modern design thinking should concentrate on designing life instead of just new artefacts.

Technology in History

The importance of technology in shaping everyday human life has been understood for a long time (Basalla 1988; Bernal 1969; Marks 1988; Pfaffenberger 1992). Many historical periods have been named after the dominant new technologies of their time (Basalla 1988), for example, the Stone, Bronze, and Iron Ages (Renfrew and Bahn 1991). Similarly, such terms as the age of steam, the age of electronics, or the age of computing are also common in characterizing the influence of specific technologies on society (Basalla 1988; Bernal 1969; Marks 1988; Pfaffenberger 1992; Williams, and Edge 1996).

Today, modern technologies (particularly information and communication technologies) are strongly, and increasingly rapidly, influencing the way we live our everyday lives (Bernal 1969; Marks 1988). Technological development has not only meant the emergence of new and more complex artefacts for people to use but it has also had a strong influence on cultures and societies.

The adoption of fire and the discovery of penicillin, for example, have significantly changed the way people live. Although the post-Homerian period in literature and the Age of Enlightenment in philosophy are vital for understanding human thinking, society, and life, no artistic innovation has had the same economic significance or concrete effect on people's everyday life as technologies have. Indeed, technical progress makes possible the development of art by creating new technical products and even new social orders (Hauser 1977). Likewise, electronic music and new media art illustrate how modern tools for producing art, such as colour or perspective techniques, can alter the type, markets, and content of art in the same way that innovations changed art during the Renaissance.

One of the key innovations of the modern age has been the steamship. It revolutionized transportation and paved the way for a new economic era (Bernal 1969; Marks 1988). It made it possible to move people, raw materials (e.g., coal), and merchandise from one place to another quickly and safely, and created reliable timetables for doing so. The new transportation technology based on steam changed companies' modes of operation and gradually altered economic life as a whole (Chandler 1992). Time became money.

Human needs have always been the trigger and justification for inventing, developing, and innovating new technologies. The demand for water is a good example. When people needed water in antiquity, complex technologies were developed to supply this need. The Romans designed aqueducts that could supply water over long distances (Chanson 2000),[1] which required a large knowledge base, from math-

---

[1] Examples from ancient times are presented in order to illustrate that the nature of many design problems has remained the same over time and in many cultures.

ematics to stone-handling practices (Carcopino 2003). In this way, the need for water became a trigger and an explanation for the construction of the aqueducts, whereas technologies provided the means to solve the problem of delivering water to the big cities. Without this technology, it would have been impossible to develop or maintain everyday life in Rome (Bruun 1991; Landels 2000). Water supply remained a challenge for scientists even in the nineteenth century. The finding that contaminated water spread illnesses led John Snow to make many inventions related to hygiene and clean drinking water (Cameron and Jones 1983), which in turn stimulated population growth and social change (Bernal 1969; Galor and Weil 2000).

Thus, technological advancements have always been intimately related to advancements in all aspects of life. Just as necessity is the mother of invention, life is the catalyst for developing technology. Technical inventions cannot be designed in a vacuum, but arise from the needs of people's everyday lives. The intimate connection between human life and technical artefacts implies that the concepts used in designing technology are gradually extended from the natural sciences to human research (Simon 1969).

Technology Emancipates People

The central function of technology can be found in the logic of making something possible or easier in order to improve the quality of life. Technologies are thus not goals in themselves, but are developed to achieve human action goals (Basalla 1988; Bernal 1969; Marks 1988). While it is possible to develop an endless number of different technologies, innovation and development processes should be based on an understanding of how new technologies can improve the quality of life. Technology can, for example, help the elderly or disabled people to carry out daily activities.

Making something happen is a form of *emancipation*, which means expanding the possibilities of life. Historically, emancipation refers to breaking free from the social conditions that enslave people (Adorno 1976; Habermas 1969, 1973; Horkheimer 1947; Stahl 2006). Life can be restricted by many kinds of necessities, difficulties, limitations, and non-ideal living conditions that prevent people from increasing their happiness. Many of these restrictions are political and social, such as the black Civil Rights or the women's liberation movement, or improving nutrition and sanitation. But many can be solved or improved through technological advancements. In medicine, for instance, new ways of treating illnesses require new kinds of technical tools, such as new medications and medical instruments.

The emancipatory role of technology has been one of the main triggers that has led many individuals and organizations to focus their efforts on creating technologies. Decreasing child mortality, illnesses, hunger, and violence, for example, has been possible with the help of technologies (Bernal 1969). While child mortality was very high 150 years ago even in developed countries, it started to rapidly decrease at the end of the nineteenth century with improvements in medical understanding, hygiene, and technology (Wolleswinkel-van den Bosch et al. 1998). Emancipation in the context of HTI thus refers to the liberation of people by technological means from any circumstances that diminish the quality of their lives.

The emancipating role of technology can best be seen in the way it shapes the world around us. For example, social discourses have been liberated by different kinds of information and communication technologies (Hansen et al. 2009; Lyytinen and Klein 1985; Ross and Chiasson 2011). In many respects, technologies have made social emancipation possible (and even necessary) as they have changed people's everyday lives. Homes, to give just one example, have undergone a technological revolution over the last 50–60 years (Cowan 1976, 1985). They have become 'industrialized' in the sense that they provide people with different kinds of technologies that take care of household tasks that were previously performed by housewives and servants. Along with household technologies, such fundamental innovations as electric lighting, industrial food production, medications, radio, phone, and television have all altered the position of women in society. Thanks to technological and social developments, women today have the opportunity to educate themselves, work outside the home and live an economically independent life (Fig. 1.1). At the same time, the number of childbirths has dramatically decreased (Subbarao and Raney 1993).

Fig. 1.1 Technology has not only changed the working methods of housework. It has also changed women's social position; today women often work outside the home and men have started to participate more in housework

New ICT technology will play a similar emancipating role in human life that the earlier forms of technologies have played in their time. While we do not know what the exact nature of this technology revolution will be, it is essential to develop our current scientific and industrial practices so they provide us with the proper tools for the coming changes. One sign of change is the political use of new media, which has made possible certain social changes—both welcome and not—which were impossible just a few decades ago (Rheingold 2012). Another interesting initiative is design by social media, for example, an opera composed recently by an international community.[2] It can be expected that more such cultural and even political changes will emerge due to innovative forms of social media and new technologies, but to make future emancipation possible, it is essential to develop our capacity to design improvements.

As can be seen from the opera example, the emancipation enabled by technology need not only concern issues of primary needs and necessities or large-scale issues in life. It can also include, for example, issues related to self-fulfilment and social freedom (Eccles and Wigfield 2002; Horkheimer 1947; Maslow 1954). For example, finding friends with similar interests and experiences on social media: that is, emancipa-

---

[2] See http://bit.se/3WiRR5

tion from loneliness. It can also mean more flexible use of one's time thanks to effective tools for distance working. Technology may improve our relationships with other people and even increase understanding between different cultures and nations. It may also make unjust social forms, such as slavery, unnecessary and unfounded. Emancipation thus means the ability to get rid of any factors that make it difficult to improve one's quality of life. Therefore, issues such as happiness, ethics, and justice should not be forgotten when technology is being developed (Bowen 2009; Jonas 1973; Stahl 2006).

As technology is social by nature, it has to be examined not only from a technical perspective but also from ethical, juridical, and political points of view (Bernal 1969; Bowen 2009; Leikas 2008, 2009; Stahl 2010). However, one should not exaggerate the influence of the logic of the possible. What is possible is not real, and what is real is only one possibility. Therefore, one cannot say that technology automatically or deterministically leads to good (Bijker 2010). It is essential to consider and design how technologies are *used*, if they are used at all. A critical example is nuclear technology: it can either be used to provide energy or to destroy all life. Thus, the mere possibility that a technology *can* do something does not in itself lead to emancipation. It is essential to return to basics and acknowledge the importance of the social and human dimensions of technology in design thinking.

The fact that improvements in technology lead to the possibility of social development (and thus social emancipation) does not mean that social emancipation will be easy. History has shown that emancipation can be a very painful process, as in the case of the emancipation of the American slaves in the nineteenth century. Nevertheless, the more we understand about the relationship between technology and social life, the more likely it is that we will use technical and technological advancement to increase the quality of life.

In sum, technologies should be designed and developed with the emancipating role of technology in mind. Technology designers should be concerned with how the technology under development will improve human life (and how people will be able to take advantage of its potential opportunities) as well as possible negative influence and harmful side effects. Thus, the goal of human emancipation can best be achieved when people's minds and lives are sufficiently understood and taken into account in the design process.

Risks of Technology

Yet technology can emancipate people only when it operates as expected. However, neither human actions nor technical artefacts are free from errors or moral problems. The use of technologies may lead to an unforeseeable accident or negative side effects, risks, or failures (Dekker 2006; Perrow 1999; Reason 1990, 1997). They may even have a disastrous influence on a society or the whole world (Beck 1992, 2008; Giddens 2000). Technologies may also entail moral risks, as they make it possible for people to harm others. Indeed, the development of technology has not been automatic and unproblematic; there have been many unwanted negative consequences and risks. For example, many natural resources, such as oil and coal, are diminishing due to the increasing need for fossil fuels to run technical inventions, and accessing them has become increasingly difficult (Aklett et al. 2010; Fantazzini et al. 2011). Another negative consequences of the growth of technology is environmental pollution; the accumulation of noble metals in the earth is becoming increasingly difficult to deal with (Anderberg et al. 2000), and nuclear accidents such as Chernobyl and Fukushima provide further examples of the risks associated with technology (Reason 2000). The prevalence of cancer has increased in those areas due to these accidents (Pflugbeil et al. 2011).

A social consequence of technology is the digital divide, which is caused by the failure in technology development to sufficiently consider democratic accessibility and the adoption of products and services (Stahl 2006, 2010). Indeed, there has been much academic discussion about the growing digital divide (Attewell 2001; Hilbert 2011; Norris 2003), technological 'haves' and 'have nots' (Howland 1998), or 'digital natives and digital immigrants' (Prensky 2001). The digital divide is no longer seen as merely an issue of access to hardware. There is now a growing concern that the lack of design foresight is creating social exclusion (Bargh and McKenna 2004). Now more than ever, the unequal adoption of (and opportunities to access) ICT excludes many from benefiting from its advantages in many fields of social life (Mancinelli 2008).

As technologies have evolved and their use has qualitatively changed, the divide is seen as separating users from non-users, and distinguishing between different types of users. There are now multiple divides that relate to a variety of factors, such as living and work conditions, ethnic background, gender, and age. Equal participation would mean that all members of society could freely use information resources to enrich their lives. Ideally, instead of being pushed aside and exploited, people from different cultures, ages, genders, physical, and mental conditions (and even educational backgrounds) could have their voices heard and their human rights recognized (Stahl 2006, 2010).

Technologies also raise many ethical questions and risks. ICT has introduced many useful things into people's everyday lives, but their use may lead to many risks with respect to information complexity, security, crime, privacy, cyber bullying, and control, to mention just a few (van Rooy and Bus 2010; Young and Quan-Haase 2009; Neves and de Oliveira Pinheiro 2010). ICT applications have, for example, provoked ethical discussion in such areas as online medical consultations and home monitoring of older people, for example, confidentiality, data protection, civil liability for medical malpractice, prevention of harm and informed consent. The development of ICT can also be threatened by faith in the omnipotence of technology. With the use of ICT technology, it is possible to instantly influence the whole world from one's home computer. The efficiency of computers, moreover, has overtaken many earlier production methods. Changes in the production of goods and services have led to changes in consuming patterns and to the creation of a technological lifestyle. Technological cultures have thus set aside traditional values and the role of values as legitimating entities for our actions. As a consequence, a justification crisis has emerged as much of the new technology lacks legitimating values (Beck 2008; Giddens 2000; von Wright 1981/2007).

Many technological products may have less visible or unintended side effects. Tobacco, for example, is a technological product in the same sense as a film or an entertaining webpage, albeit a much more problematic one. Smoking leads to addiction and serious health risks.[3] Thus, manufacturing and selling the product is ethically questionable, and attempts have been made to restrict its use via warnings and legislation. Sometimes great benefits create huge risks. Such substantial contributors to health as improved hygiene and advanced medication and vaccination are also major explanatory factors for uncontrolled population growth, which has caused social problems in many places around the world. For example, improved female education and social changes have practically stopped overall world population growth (Lutz et al. 2001). And although the number of cars has increased, improved cars, traffic conditions, and speed limits have reduced the number of fatal traffic accidents. Such notions as a *risk society* can characterize imminent societal risks (Beck 2008).

The notion of a risk society includes the understanding that risks are global because modern technological societies are global (Beck 2008). Risks arise because it is difficult to predict the consequences of technological developments such as improved communications. Modern ICT opens up possibilities for people to act, but also contributes to the emergence of different threats. For example, it creates new opportunities for schoolchildren to proceed in their education, hobbies, and social life, but it may also help foster a small marginal group of school bullies, and even school killers and terrorists with dangerous technological weapons (Beck 2008). Improving the technological solutions in question or responding to problems via proper social means can mostly be used to eliminate problems caused by new technologies. It is hardly realistic to eliminate technical problems by abandoning technologies. The paper industry, for example, has produced a great deal of environmental waste for over a century, but giving up the use of paper is hardly a viable solution. Instead, developing closed processes that produce less waste and can effectively reuse the waste materials is a much more rational solution (O'Brien and O'Brien 1978). Thus, technology can also be used to solve the problems it causes.

Technologies themselves do not cause accidents. The risks associated with the use of new technologies are caused by human error during the processes of design, manufacturing, or use (Perrow 1999; Reason 1990,

---

[3] www.cdc.cov.tobacco

2000). It is therefore important to understand the logic of the emergency of risks. Research must be able to uncover why some technologies are prone to generating different kinds of risks (in addition to providing positive outputs) so that designers can find the means to eliminate them. This kind of research requires a proper understanding of the human dimension of technology. Technologies comprise a combination of technical artefacts, such as programs, machines, devices or tools, as well as their uses by human beings for a specific purpose (Meier-Oeser 1998). Therefore, technical artefacts should be designed with the human application in mind. In this sense, designers should concentrate not only on designing artefacts but also on designing human action. Ultimately, designing technologies includes designing how people can (and do) live and improve their lives.

Towards a New Interaction Design Culture

The paradoxical capability of technology design to create both opportunities and threats indicates that it is necessary to consider technology design and development from the perspective of humans as well as technical artefacts. Thus, well-considered facts of human research should be introduced into technology design thinking.

In today's engineering, basic sciences such as mathematics, physics, and chemistry provide conceptual and empirical tools that make design possible (Adams 1992; Dym and Brown 2012). History has shown that science is a prerequisite of modern innovations. Einstein, for example, understood the intimate connection between matter and energy. He also realized that it was possible to create a chain reaction when newly freed electrons meet new nuclei. This theoretical phenomenon was later applied by Fermi in nuclear power stations and Oppenheimer's group in creating the atom bomb. Here, basic natural science created tools for very complicated engineering thinking and applications. In the same way, social sciences can create important visions for political and social thinking.

The sciences have not always directly influenced technology creation. For example, Henry Ford was a former blacksmith and clock smith who had a sufficient degree of curiosity, creativity, and independent thinking to create the car industry, one of the world's most important (Ford 2009). However, today it would be difficult to imagine a car company that does not use scientific knowledge on materials, mechanics, and chemistry, or marketing research and ergonomics. When the car industry was in its infancy, the amount of research required was modest. Today, all kinds of research expertise are required to solve design problems.

Traditionally, as in the case of the car industry, developing any technological device has involved shaping the 'steel' or changing mechanical aspects of the artefact. Yet today it is possible to create new technologies by making changes in the software. The core question has changed from the physical structures of the technologies to how well they can be integrated into people's everyday lives. This leads to a form of design thinking in which human concepts become central.

A good example of modern developments is social media, which has diverse services for which Facebook-like applications have paved the way. With the help of social media, it has been possible to create new communication possibilities for all kinds of user segments without having to make any technical changes to the artefacts. Social and other types of new media services such as blocks, wikis, groups, or podcasts give even users with minimal technical understanding the opportunity to create new services and adopt new ways of using technologies to reach their goals. Consequently, understanding people becomes more central when designing these kinds of services.

Digital and business ecosystems are another example of new technology design types. They are open socio-technical systems, and are often clusters of enterprises that have networked into effective business systems (Popp and Meyer 2010). Digital ecosystems can also involve teaching or entertainment networks, or any self-organizing net communities that can organize new social uses for technologies. In these ecosystems, the contributing contents of new technologies is most important—that is, they depend more on people and their everyday life situations than on the technical artefact itself.

Changes in the nature and role of technical artefacts in people's lives also make it necessary to rethink the foundations of HTI thinking. What is the role of human knowledge in modern interaction design? What form should the relevant knowledge take, and how should it be used in practical design work? What are the truly important questions, and how should they be answered?

Only a small part of HTI design solutions can be based on knowledge of mathematics, physics, or chemistry. It relies more heavily on different forms of human research, from physiology to psychology and sociology, to help understand how knowledge of the human mind and life should be incorporated into design solutions (Card et al. 1983; Carroll 2003; Karwowski 2006; Rosson and Carroll 2002). The long research tradition in the human and social sciences can offer concepts, methods, and empirical theories to carry out HTI design on rational and scientific grounds. The core question is how this knowledge can be effectively used to create new technological innovations. So far, usability tests for products and prototypes have been the most common form of applying human research. Using this approach, design relies on common sense guidelines and design solutions that are systematically tested and tried out (Cooper et al. 2007; Dumas and Redish 1999; Shneiderman and Plaisant 2005). This is good practice, but products can easily fail if human research knowledge and concepts are limited to usability issues and are not applied throughout the design and development process.

In HTI design, the design ideas should be based on knowledge of the human mind in the same way that the idea of the atomic bomb was based on the latest developments in nuclear physics. Understanding the human dimension may lead to important, practical technological solutions. It may also avoid unnecessarily expensive technological solutions. Steve Jobs' innovations, for example, were much more related to what people do in life than to technical issues. The potential applications of human research (and the ways in which it can be used in technology design) require metascientific investigation.

One way to explain the necessity of applying the concepts of human life to produce successful design outcomes is to open up the basic metascientific difference between the logics of the causal natural sciences and intentional human research (Bunge 1959; Radnitsky 1968; Stegmüller 1969; von Wright 1971). In causal explanations, the phenomenon that should be explained (i.e., the explanandum) follows the phenomenon that explains (the explanans), but in intentional or teleological explanations, something that happens after the event explains the explanandum (von Wright 1971).

The development of products or services should take into account what people will achieve by adopting the technology and how their lives will change. People use an excavator to remove soil when building a house. Earth removal precedes the construction of a house, so building a house cannot be the *cause* of using the excavator, though it provides an *explanation* of its use. To understand human actions, one has to be able to explain the reasons behind the deeds; these reasons are the goals of the eventual actions.

It is thus essential to investigate all the roles that human knowledge plays in HTI design and development. This presupposes the creation of a new technology design culture in which knowledge of human performance, the mind, and different forms of social life will be given a more central role. It also presupposes a metascientific analysis of HTI design thinking and culture. As a change is taking place already, it is time to consider what has been achieved so far in developing the human dimension of technology design in order to accurately define the role of human research and the modes of its implementation in HTI design.

Human Turn

The fast development of ICT, the multifunctionality of devices, and future emerging technologies such as sensors and robots challenge us to discuss the future scenario from the point of view of human beings. New ICT issues, such as embeddedness and mobility and the presence of technology in all aspects of our life, are changing our relationship to the environment, to other people and even with ourselves.

Although technologies have always been targeted at people—that is, to serve and improve people's lives in some way—there has been little systematic use of human research knowledge in designing technologies. The main interest in design has been how to create functional technical artefacts, which has been based on knowledge of mathematics, physics, and chemistry as well as the systematic use of the laws of nature (Pahl et al. 2007).

Today, as technologies have become increasingly complex, designers have been forced to focus more and more on ensuring that users really can use the technical artefacts. Ease of use, usefulness and ease of adoption, together with trust, have been found to be important elements of user satisfaction and acceptance (Davis 1989; Venkatesh 2000). The increased attention to these basic design questions is not surprising, given the countless ICT products that have been unable to interest or satisfy potential users. Too often, products have failed due to poor design in terms of usability, risk and error prevention, adoption, and innovation.

Throughout history, technology has been a catalyst for human social development. To make the most of new technological inventions, designers should turn their attention to how their ideas affect people's lives. They should acknowledge that they are not merely designing artefacts but instead different ways for people to live their everyday lives. The gradually increasing importance of human research in technology design became more obvious during the Second World War, when such research areas as ergonomics and human factors were introduced (Karwowski 2006). With computers and computing, human research has become even more important than with mechanical technologies. Since user interfaces have become more complex, usage issues are now an important challenge for designers. Today there is widespread acknowledgement that technology should be integrated into everyday life. The gradual refocusing in design thinking has involved a new 'human turn'. It cannot replace classical technical design, but it can refocus design thinking as a whole. Since technology is to be used by people, it is essential to consider how the new products and services should be integrated into their lives.

References

Adams, J. L. (1992). *Flying buttresses, entropy, O-rings: The world of engineer.* Cambridge, MA: Harvard University Press.

Adorno, T. H. (1976). *The positive dispute in German sociology.* London: Heinemann.

Aklett, K., Höök, M., Jacobsson, K., Lardelli, M., Snowden, S., & Söderberg, B. (2010). The peak oil age—Analyzing the world oil production reference scenario in world energy outlook. *Energy Policy, 38,* 1398–1414.

Anderberg, S., Prieler, S., Olendrzynski, K., & de Bruyn, S. (2000). *Old sins.* Tokyo: United Nations University Press.

Attewell, P. (2001). Comment: The first and second digital divides. *Sociology of Education, 74,* 252–259.

Bargh, J. A., & McKenna, K. Y. (2004). The internet and social life. *Annual Review Psychology, 55,* 573–590.

Basalla, G. (1988). *The evolution of technology.* Cambridge: Cambridge University Press.

Beck, U. (1992). *Risk society: Towards a new modernity.* London: Sage.

Beck, U. (2008). *Weltrisikogesellschaft* [World risk society]. Frankfurth am Main: Surkamp.

Bernal, J. D. (1969). *Science in history.* Harmondsworth: Penguin.

Bijker, W. E. (2010). How is technology made?—That is the question! *Cambridge Journal of Economics, 34,* 63–76.

Bowen, W. R. (2009). *Engineering ethics: Outline of an aspirational approach.* London: Springer.

Bruun, C. (1991). *The water supply of ancient Rome: A study of Roman imperial administration.* Helsinki: Societas Scientiarum Fennica.

Bunge, M. (1959). *Causality and modern science.* New York: Dover.

Cameron, D., & Jones, I. G. (1983). Johns snow, the broad street pump and modern epidemiology. *International Journal of Epidemiology, 12,* 393–396.

Carcopino, J. M. (2003). *Daily life in ancient Rome: The people and the city at the height of the empire.* London: Routledge and Sons.

Card, S., Moran, T., & Newell, A. (1983). *The psychology of human-computer interaction.* Hillsdale, NJ: Erlbaum.

Carroll, J. M. (1997). Human computer interaction: Psychology as science of design. *Annual Review of Psychology, 48,* 61–83.

Carroll, J. M. (Ed.). (2003). *HCI models, theories, and frameworks: Toward a multidisciplinary science.* San Francisco, CA: Morgan Kaufmann.

Chandler, A. D. (1992). Organizational capabilities and the economic-history of the industrial-enterprise. *Journal of Economic Perspectives, 6,* 79–100.

Chanson, H. (2000). Hydraulics of roman aqueducts: Steep chutes, cascades, and dropshafts. *American Journal of Archaeology, 104,* 47–72.

Collison, S., Budd, A., & Moll, C. (2009). *CSS mastery: Advanced web standards solution.* Berkeley, CA: Friends of ED.

Cooper, A., Reimann, R., & Cronin, D. (2007). *About Face 3: The essentials of interaction design.* Indianapolis, IN: Wiley.

Cowan, R. S. (1976). The 'industrial revolution' in the home: Household technology and social change in the 20th century. *Technology and Culture, 17,* 1–23.

Cowan, R. S. (1985). The industrial revolution in the home. In D. Mackenzie & J. Wajcman (Eds.), *The social shaping of technology* (pp. 181–201). London: Taylor and Francis.

Davis, F. D. (1989). Perceived usefulness, perceived ease of use, and user acceptance of information technology. *MIS Quarterly, 13*, 319–340.

Davis, M. C., Challenger, R., Jayewardene, D. N., & Clegg, C. W. (2014). Advancing socio-technical systems thinking: A call for bravery. *Applied Ergonomics, 45*, 171–180.

Dekker, S. (2006). *The field guide to understanding human error*. Farnham: Ashgate.

Dumas, J. S., & Redish, J. (1999). *A practical guide to usability testing*. Exeter: Intellect Books.

Dym, C. L., & Brown, D. C. (2012). *Engineering design: Representation and reasoning*. New York: Cambridge University Press.

Eccles, J., & Wigfield, A. (2002). Motivational beliefs, values, and goals. *Annual Review of Psychology, 53*, 109–132.

Fantazzini, D., Höök, M., & Angelantoni, A. (2011). Global oil risks in the early 21st century. *Energy Policy, 39*, 7865–7873.

Ford, H. (2009). *My life and my work*. New York: Classic House.

Galor, O., & Weil, D. (2000). Population, technology, and growth: From Malthusian stagnation to the demographic transition and beyond. *American Economic Review, 90*, 806–828.

Giddens, A. (2000). *Runaway world: How globalization is reshaping our lives*. Cambridge: Polity Press.

Goodwin, K. (2011). *Designing for the digital age: How to create human-centered products and services*. Indianapolis, IN: Wiley.

Habermas, J. (1969). *Technik and Wisseschaft als 'ideologie'* [Technology and science as ideology]. Frankfurth am Main: Surkamp.

Habermas, J. (1973). *Erkentniss und interesse* [Knowledge and interests]. Frankfurth am Main: Surkamp.

Hansen, S., Berente, N., & Lyytinen, K. (2009). Wikipedia, critical social theory, and the possibility of rational discourse 1. *The Information Society, 25*, 38–59.

Hassenzahl, M., & Tractinsky, N. (2006). User experience—A research agenda. *Behaviour and Information Technology, 25*, 91–97.

Hauser, A. (1977). *The social history of art* (Vols. 1–3). London: Routledge.

Helander, M., & Khalid, H. M. (2006). Affective and pleasurable design. In G. Salvendy (Ed.), *Handbook of human factors and ergonomics* (pp. 543–572). Hoboken, NJ: Wiley.

Hilbert, M. (2011). The end justifies the definition: The manifold outlooks on the digital divide and their practical usefulness for policy-making. *Telecommunications Policy, 35,* 715–736.

Horkheimer, M. (1947). *The eclipse of reason.* New York: Oxford University Press.

Houkes, W., Kroes, P., Meijers, A., & Vermaas, P. E. (2011). Dual-nature and collectivist frameworks for technical artefacts: A constructive comparison. *Studies in History and Philosophy of Science, 42,* 198–205.

Howland, J. S. (1998). The 'digital divide': Are we becoming a world of technological 'haves' and 'have-nots?'. *Electronic Library, 16,* 287–289.

International Organization for Standardization. (1998a). *ISO 9241-11: Ergonomic Requirements for Office Work with Visual Display Terminals (VDTs): Part 11: Guidance on Usability.*

International Organization for Standardization. (1998b). *ISO-14915: Ergonomic Requirements for Office Work with Visual Display Terminals (VDTs): Part 11: Guidance on Usability.*

Jonas, H. (1973). Technology, responsibility: Reflections on the new task ethics. *Social Research, 40,* 31–54.

Karwowski, W. (2006). The discipline of ergonomics and human factors. In G. Salvendy (Ed.), *Handbook of human factors and ergonomics* (pp. 3–31). Hoboken, NJ: Wiley.

Kim, K., Jacko, J., & Salvendy, G. (2011). Menu design for computers and cell phones: Review and reappraisal. *International Journal of Human–Computer Interaction, 2,* 383–404.

Landels, J. G. (2000). *Engineering in the ancient world.* Berkley, CA: University of California Press.

Leikas, J. (2008). *Ikääntyvät, teknologia ja etiikka—näkökulmia ihmisen ja teknologian vuorovaikutustutkimukseen ja—suunnitteluun* [Ageing, technology and ethics—views on research and design of human-technology interaction] (VTT Working Papers No. 110). Espoo: VTT.

Leikas, J. (2009). *Life-based design—A holistic approach to designing human-technology interaction.* Helsinki: Edita Prima Oy.

Lutz, W., Sanderson, W., & Scherbov, S. (2001). The end of world population growth. *Nature, 412,* 543–545.

Lyytinen, K., & Klein, H. K. (1985). The critical theory of Jurgen Habermas as a basis for a theory of information systems. In E. Mumford, R. Hirschheim, G. Fitzgerald, & A. T. Woods-Harper (Eds.), *Research methods in information systems* (pp. 219–236). New York: North Holland.

Mancinelli, E. (2008). E-inclusion in the information society. In R. Pinter (Ed.), *Information society: From theory to political practice: Course book*. Budapest: Gondolt–Új Mandátum.

Marks, J. (1988). *Science in the making of the modern world*. London: Heinemann.

Maslow, A. H. (1954). *Motivation and personality*. Oxford: Harpers & Row.

Meier-Oeser, S. (1998). Technologie [Technology]. In J. Ritter & K. Gründer (Eds.), *Historisches Wörterbuch der Philosophie* (Vol. 10, pp. 958–961). Darmstadt: Wissenschaftliche Buchgesellschaft.

Nagamashi, M. (2011). Kansei/affective engineering and history of Kansei/affective engineering in the world. In M. Nagamashi (Ed.), *Kansei/affective engineering* (pp. 1–30). Boca Raton, FL: CRC Press.

Neves, J., & de Oliveira Pinheiro, L. (2010). Cyberbullying: A sociological approach. *International Journal of Technoethics, 1*, 24–34.

Nielsen, J. (1993). *Usability engineering*. San Diego, CA: Academic Press.

Norman, D. A. (2002). *The design of everyday things*. New York: Basic Books.

Norris, P. (2003). *Digital divide: Civic engagement, information poverty, and the internet worldwide*. Oxford: Taylor and Francis.

O'Brien, R. C., & O'Brien, E. R. (1978). *Method for recycling paper mill waste water*. United States Patent 4,115,188.

Orlikowski, W. J. (1991). Duality of technology: Rethinking the concept of technology in organizations. *Organization Science, 3*, 398–427.

Orlikowski, W. J. (2000). Using technology and constituting structures: A practical lens for studying technology in organizations. *Organization Science, 11*, 404–428.

Pahl, G., Beitz, W., Feldhusen, J., & Grote, K. H. (2007). *Engineering design: A systematic approach*. Berlin: Springer.

Perrow, C. (1999). *Normal accidents: Living with high-risk technologies*. Princeton, NJ: Princeton University Press.

Pfaffenberger, B. (1992). Social anthropology of technology. *Annual Review of Anthropology, 21*, 491–516.

Pflugbeil, S., Paulitz, H., Claussen, A., & Schmitz-Feuerhake, I. (2011). *Health effects of Chernobyl: 25 years after the reactor catastrophe*. Berlin: Gesellschaft fuer Strahlenschutz.

Popp, K., & Meyer, R. (2010). *Profit from software ecosystems*. BoD–Books on Demand.

Prensky, M. (2001). Digital natives, digital immigrants part 1. *On the Horizon, 9*, 1–6.

Radnitsky, G. (1968). *Contemporary schools of metascience.* Göteborg: Akademieförlaget.

Reason, J. (1990). *Human error.* Cambridge: Cambridge University Press.

Reason, J. T. (1997). *Managing the risks of organizational accidents.* Aldershot: Ashgate.

Reason, J. (2000). Human error; models and management. *British Journal of Medicine, 320,* 768–770.

Renfrew, C., & Bahn, P. G. (1991). *Archaeology: Theories, methods, and practice.* London: Thames and Hudson.

Rheingold, H. (2012). *Smart mobs.* Cambridge, MA: MIT Press.

Ross, A., & Chiasson, M. (2011). Habermas and information systems research: New directions. *Information and Organization, 2,* 123–141.

Rosson, B., & Carroll, J. (2002). *Usability engineering: Scenario-based development of human-computer interaction.* San Francisco, CA: Morgan Kaufmann.

Saariluoma, P., Nevala, K., & Karvinen, M. (2006). Content-based analysis of modes in design engineering. In J. Gero & A. Goel (Eds.), *Design computing and cognition '06* (pp. 325–344). Springer: Berlin.

Shneiderman, B., & Plaisant, C. (2005). *Designing user interfaces.* Boston, MA: Pearson.

Simon, H. A. (1969). *The sciences of artificial.* Cambridge, MA: MIT Press.

Stahl, B. C. (2006). Emancipation in cross-cultural IS research: The fine line between relativism and dictatorship of intellectual. *Ethics and Information Technology, 8,* 97–108.

Stahl, B. C. (2010). 6. Social issues in computer ethics. In L. Floridi (Ed.), *The Cambridge handbook of information and computer ethics* (pp. 101–115). Cambridge: Cambridge University Press.

Stegmüller, W. (1969). *Hauptströmungen der gegenwartsphilosophie: Eine kritische einführung* [Main traditions of modern philosophy]. Stuttgart: Kröner.

Subbarao, K., & Raney, L. (1993). *Social gains from female education.* Washington, DC: World Bank.

van Rooy, D., & Bus, J. (2010). Trust and privacy in the future internet—A research perspective. *Identity in the Information Society, 3,* 397–404.

Venkatesh, V. (2000). Determinants of perceived ease of use: Integrating control, intrinsic motivation, and emotion into the technology acceptance model. *Information Systems Research, 11,* 342–365.

von Wright, G. H. (1971). *Explanation and understanding.* London: Routledge and Kegan Paul.

von Wright, G. H. (1981/2007). *Humanismi elämänasenteena* [Humanistic stand to life]. Helsinki: Otava.

Williams, R., & Edge, D. (1996). The social shaping of technology. *Research Policy, 25*, 865–899.

Wixon, D., & Wilson, C. (1997). The usability engineering framework for product design and evaluation. In J. Jacko & A. Sears (Eds.), *Handbook of human-computer interaction* (pp. 653–668). Amsterdam: Elsevier Science.

Wolleswinkel-van den Bosch, J. H., van Poppel, F. W., Tabeau, E., & Mackenbach, J. P. (1998). Mortality decline in the Netherlands in the period 1850–1992: A turning point analysis. *Social Science and Medicine, 47*, 429–436.

Young, A. L., & Quan-Haase, A. (2009). Information revelation and internet privacy concerns on social network sites: A case study of Facebook. In: *Proceedings of the Fourth International Conference on Communities and Technologies* (pp. 265–274).

2

Design Discourses

A newcomer in the human-technology interaction (HTI) field encounters a complex variety of interest groups and design discourses. Over the years, human interaction with artefacts has been studied from many different approaches, perspectives, concepts, and methodologies. Yet conceptual structuring of the field is much needed; traditional metascientific concepts such as paradigm, research programme, and discourse can help.

Ultimately, designing is thinking. Therefore, it should be studied as *human* thinking (Cross 1982, 2001, 2004; Simon 1969). In design processes, technical or art designers pursue the creation of something new, while thinking creates new pieces of information. These processes are typical and unique to the human mind. Apes can use primitive 'tools' and 'symbols', but they do not have language, much less the Internet, mass production, or a worldwide economy (Deacon 1997). All HTI design is unified by the fact that it is a form of human thinking (Cross 1982, 2004; Dym and Brown 2012; Simon 1969).

However, interaction design thinking is not a unified and monolithic whole; human social thinking seldom is. Modern HTI designers have

© The Editor(s) (if applicable) and The Author(s) 2016   **25**
P. Saariluoma et al., *Designing for Life*,
DOI 10.1057/978-1-137-53047-9_2

developed a variety of different ways of thinking about (and approaches to) HTI design. In these, such concepts as HCI, ergonomics, action theory, and cognitive modelling often demarcate different designers (Saariluoma and Oulasvirta 2010). Yet it is problematic that these different approaches use different concepts, ask different questions, and rely on different truths. To bring clarity to the field, and find out how design thinking is organized as a whole, one can apply the major outcomes of modern metascientific research: paradigm, research programme, and discourse (Kuhn 1962; Lakatos 1970; Laudan 1977).

Human thinking is essentially about asking questions, setting problems, and solving them (Newell and Simon 1972). However, before one can ask a question (or answer it), one has to be able to formulate the question both conceptually and in verbal sentences that can be communicated to other people. Thus, it is necessary to pay attention to the concepts that researchers use in their different approaches to HTI design problems, including credos, truths, goals, and focus areas. Such analysis of the foundations of the common ways of thinking is essential for creating explicit and tacit foundations for technological thinking (Saariluoma 1997).

How Researchers and Designers Perceive HTI

HTI design and research is dominated by different groups of researchers that are organized around certain common ideas. Some of them are specialized in, for example, style guides and usability, others in user experience; some are keen on universal or inclusive design, while yet others strive for pleasurable interaction. There are many interest groups working on HTI issues and related topics.

Some of the topics and interest groups are so closely related that it can be somewhat difficult to distinguish between them. For example, differentiating between usability and user experience (Nielsen and Norman 2014), and Kansei engineering[1] (Nagamashi 2011) and user

---

[1] This type of engineering uses semantic differential-based methods to investigate people's preferences for various products.

experience (Hassenzahl 2011) has proven to be problematic. Further, the basic premise of Kansei engineering is shared by 'affective ergonomics'; they both focus on emotional processes in HTI (Helander and Khalid 2006; Nagamashi 2011). Thus, there is considerable overlap in current discourses.

The field of HTI therefore involves social groups of researchers and designers aiming to achieve the same goals from different perspectives. Creating a deeper understanding of the field requires synthesizing (or at least acknowledging the existence of) the various schools of interaction design and research thinking.

Many interesting approaches influence the field of HTI. For example, many new innovations have been made in the field of geron-technology, which focuses on designing technologies to support the everyday life of ageing people (Bouma et al. 2009). Likewise, funology and entertainment computing, which focus on creating technologies such as games that bring joy to the users, has had an impact on HTI (Monk et al. 2002; Rauterberg 2004, 2010). HCI is focused on the interaction between humans and computers. A similar approach is ergonomics, which focuses on the mechanical control of machines. These two approaches are converging with the increasing prevalence of computer technology in machines; the difference between them is becoming irrelevant.

The different approaches examine the field of HTI from slightly different points of view and open up important perspectives in HTI design. This is important in extending rational thinking in designing interaction. While none of the approaches is insignificant, none provides an exhaustive coverage of the entire field. Likewise, individuals, researchers, and designers have different life and work experiences, and thus they bring different intuitive assumptions to science (Saariluoma 1997) and a number of slightly different approaches to HTI, which Kuhn (1962) calls paradigms or models of thought. Is it possible to understand the field of HTI research and design in terms of competing paradigms? To answer this question, one has to explore the definition and characteristics of a paradigm in research and design.

Anatomy of Paradigms

Paradigms are based on scientific achievements that have become ideals or models of thought for other researchers (Kuhn 1962). They are created when a community of practitioners (including researchers and designers) defines problems, concepts, credos, procedures, and even acceptable facts within certain frameworks (Kuhn 1962: viii). Some paradigms are extensive, such as Galileo's experimental natural science, while some are much narrower, such as the working memory paradigm in psychology (Baddeley 1986). In all cases, they are approaches adopted by a number of researchers or designers who share an ideal model of thinking and shaping their work.

One can find fields of research and design that diversify paradigms in all fields of learning (Kuhn 1962). One could compare a paradigm with a line of inquiry in a detective novel. For example, when a detective focuses only on the blond lady and asks questions concerning this particular individual, the true criminal (the middle-aged barber) may be ignored and forgotten. Yet while the lines of inquiry are different—and different issues become important in them—both lines might be needed to help the detective solve the crime (Hanson 1958).

In every field of professional work, good conceptual organization is essential for the work to be rich and productive. In art and architecture, Scandinavian design, for example, is a different paradigm from older styles (Alexander 1977; Koivisto 2011; Saarinen and Saarinen 1962; Whitford and Ter-Sarkissian 1984). In technical design, touchscreen technology, what you see is what you get (WYSIWYG), or direct manipulation define their own spheres of thinking (Galiz 2002; Shneiderman and Plaisant 2005; Shneiderman and Maes 1997), while in programming, structural or object-oriented programming paradigms are commonly used (Budd 1991; Dijkstra 1972).

Paradigms define the topic of research, the concepts researchers use, and the methods that are considered suitable (Barbour 1980; Kuhn 1962). They also detail the questions that are legitimately asked, the types of explanations sought, and the types of solutions that are seen as acceptable (Barbour 1980; Kuhn 1962). Paradigms also define the implicit assumptions about what the world is like that is shared by their adherents

(Kuhn 1962; Saariluoma 1997). They also entail knowledge about how to apply the findings in practice. From a sociological point of view, paradigms provide information about the suitable outlets for research and the recognized congresses and events. They even contribute to creating 'gurus' in the field, whose opinions carry significant weight within the paradigm.

In sum, a paradigm specifies:

* topic and concepts of research;
* intuitive assumptions concerning the topic of research;
* legitimate questions;
* structure of questions and nature of legitimate answers;
* suitable methods and instrumentations; and
* social organization of the research.

Paradigms outline the ways in which research is conducted in a field of science or design. They help researchers generate ideas, but they do not—and should not—define the absolute limits of what can (or should) be studied (Feyerabend 1975). Researchers should not slavishly follow all the assumptions of a certain paradigm, as many concepts are seen differently by the researchers and, depending on the particular studies that have been conducted, the importance of the competing concepts and methods may vary. Nevertheless, this does not mean that it would be impossible to outline an anatomy of a paradigm and use it to compare different paradigms. Finally, paradigms are not eternal. They are created at a point in time, live in a particular time, and disappear when they are rejected or replaced by a more intelligent way of investigating the problem (Kuhn 1962). The current HTI approaches include elements of a paradigm, which is a useful way to consider and define what it means to follow a particular approach.

Main Paradigms of HTI Design: An Overview

In order to define the paradigmatic structure of HTI research, some of its major paradigms should be outlined. This section discusses the most important HTI paradigms in more or less historical order in order to provide an overview of how HTI research and design are conducted today.

HTI issues have been traditionally discussed in technical concepts, which mean that researchers and developers focus on existing documentation or a variety of technical standards for measuring and controlling machines and devices (Schmidt and Lee 2011; ISO 1998). Here, the main goal has been to create technical paradigms that facilitate rational interaction between technology and people. A good example of an interaction paradigm that enables developers to build more effective interaction processes is the development of programming. Game developers, for instance, used to build their games from scratch. Today there are numerous game engines available, which allow them to focus more on the content and artistic designs (Lewis and Jacobson 2002).

Many HTI issues have their origins in technical problems. For example, the problem of friction on the surface of a road is a central safety issue, which concerns both the tyre industry and road designers. It is a problem with great value in making good HTI, but human research plays a minor role in this discussion. Another example of indirect HTI intervention is the technical ontologies of information systems. Internal ontological structures of information systems—for example, concerning traffic, construction, or healthcare services—are often based on functionalities of the system that have to be considered from a technical point of view. These, in turn, are directly linked with human users and usage situations (Niezen 2012).

Human factors and ergonomics form two original, practically identical paradigms, in which human research concepts are vital. Karwowski (2006) dates the beginning of ergonomics to the mid-nineteenth century, but it could easily be placed in Ancient Greece or even earlier. The original field of research of human factors and ergonomics was the interaction between people and machines and the factors that affect that interaction (Bridger 2009). The research was greatly boosted by the Second World War (Wickens and Holands 2000). The focus has been on fostering effective performance and eliminating harmful effects. The key problems in this paradigm have been physiological and cognitive, but recently such broader issues as affective, organizational, and cross-cultural interaction have also become fashionable (Aykin et al. 2006; Helander and Khalid 2006).

A closely related paradigm to human factors and ergonomics is HCI, which began much later (Dix et al. 1993; Preece et al. 1994, 2004), which has long focused on behavioural goals and user tasks in work settings. HCI can be thought of as a sub-area of ergonomics, but many of the problems of ergonomics and HCI are quite different and the two fields should be seen as different paradigms. For example, not all physical objects have a computational dimension: people interact with buildings, walls, roads, tableware, and saucepans, for example, and although these items might have been designed with the help of computer software, they do not include a direct interaction with a computer.

It can be argued that the beginnings of the field of HCI can be traced to a Gaithersburg conference on human factors in computing in 1982, early user modelling (Card et al. 1983; Cockton 1987), or the publication of Weinberg's (1971) book *The Psychology of Computer Programming* which was intended to provide tools to address the so-called software crises, that is, the fact that computers offered more possibilities than the programmers of the time could utilize. These two events show how the HCI paradigm grew independently of human factors and ergonomics as computers and their use had gained significant practical value.

The main question of HCI is how people and computers influence each other (Dix et al. 1993). It is built around three major concepts: people, computers, and interaction, each of which must be interpreted flexibly. Computers, for example, can be embedded inside numerous solutions; interaction can be direct or indirect, text based, graphical, or ubiquitous; and people can be analysed using many different conceptual systems such as limited capacity, culture, or learning.

Methodologically, HCI has been dominated by two important approaches: design and evaluation (Coursaris and Bontis 2012; Dix et al. 1993; Preece et al. 1994). Design methods such as paper and rapid prototyping, task analysis, contextual inquiry, and scenario-based design have been used to create new interaction processes. Evaluation methods such as heuristic evaluation, experimentation, cognitive and pluralistic walkthroughs, and field trials should ensure that the real interfaces will operate as smoothly as possible (Markopoulos and Bekker 2003; Nielsen 1993).

The HCI research paradigm has generated a number of important truths that have provided a structure for the field. For example, cognitive complexity theory (Anderson et al. 1984; Dix et al. 1993; Kieras and Meyer 1997; Newell and Simon 1972) is based on the limited capacity of the human mind, which was originally found by Miller (1956) in memory and Broadbent (1958) in attention. In a nutshell, the theory asserts that the more complex an interface or task is, the slower and more error prone is human performance. Practically all fields of interaction research, such as ergonomics, human factors, or HCI, recognize the meaning of limited capacity as well as the explanatory role of complexity (Covan 2000; Gopher and Donchin 1986; Di Stasi et al. 2013). Prominent HCI researchers include Norman, Card, and Moran. Its best-known conferences are the ACM SIGCHI, HCI International, and Interact, and its key journals are *Human-Computer Interaction, ACM interactions, Interacting with Computers,* and the *International Journal of Human-Computer Studies.*

The role of emotions in design thinking and design research has been recognized for centuries. In the early 1960s, Japanese researchers developed Kansei engineering in order to design affective components (Nagamashi 2011). Likewise, user experience (UX) research, which has gained ground with the growth of interactive products and services, focuses on the subjective, situated, and positive aspects of technology use. It was initially developed by Norman et al. (1995), but many active researchers have since contributed to it (Hassenzahl and Tractinsky 2006; Norman 2004; Preece et al. 2004; McCarthy and Wright 2004). These researchers have emphasized the importance of human experience in defining design goals. Thus, the basic goal of the UX research paradigm is to design for pleasure: that is, to create positive and pleasant (even aesthetic) interaction conditions in which users feel good.

The basic concept of UX is experience (especially emotional experience). Naturally, technology is also central, but the research concentrates on the *uses* of technology, such as services, social interaction, and gaming (Forlizzi and Battarbee 2004; Väänänen-Vainio-Mattila et al. 2009; McCarthy and Wright 2004). UX also reaches beyond the concept of 'computer', as all products and services involve a psychological experience (Saariluoma and Jokinen 2014).

Other paradigms have similar goals to Kansei engineering and UX. They also emphasize such issues as pleasure, hedonic values, flow, competence, and frustration (Saariluoma and Jokinen 2014; Seligman and Csikszentmihalyi 2000). Typical examples of such paradigms are empathic design, funology, entertainment computing, and emotional user psychology (Mattelmäki and Batterbee 2002; Monk et al. 2002, Rauterberg 2010; Saariluoma and Jokinen 2014). One might argue that Kansei engineering and UX research, as well as emphatic design, could be considered different variants of affective ergonomics. Kansei, however, has a well-defined methodology based on semantic differentials, while UX researchers use a somewhat broad methodological approach (Nagamashi 2011), and the communities that support the two approaches are different.

Some design approaches seek the users' perspective in the early phases of product or service design in order to design for quality of life. Typical examples of these are such *value-oriented design* paradigms as computer ethics, value-sensitive design, and worth-centred development, which consider values from different perspectives. The former two are interested in ethical values and especially information (computer) ethics, which deals with ethical problems that are created, transformed, or exacerbated by ICT (Bowen 2009; Cockton 2004), whereas the latter is interested in technology's role in value creation in general (Cockton 2004). Related paradigms include empathic design, inclusive design, and even geron-technology. The focus of ethical paradigms is not only on immediate interaction with artefacts but also on much broader issues of the social, ethical, and environmental impacts of technology as well as responsible research and innovation (Stahl 2006, 2010; van Schomberg 2013).

The relationship of technology to some aspects of life is also considered in socio-technical research, activity theory, cross-cultural ergonomics, and different variations of task analysis (Annett 2000, 2004; Kaptelinin 1996; Kuutti 1996; Vygotsky 1980). In these paradigms, researchers mainly consider implicitly issues, which define why people use technologies at all (Abras et al. 2004; Annett 2000, 2004; Annett and Duncan 1967; Card et al. 1983; Karwowski 2006; Kuutti 1996; Mao et al. 2005; Marti and Bannon 2009; Nardi 1996; Norman and Draper 1986; Stahl 2006, 2010). They inspect the role of technology in society (particularly the social role) and how its use should be integrated with what people do. The instrumentation and methods of these paradigms are close to those typical to the social sciences.

This brief overview of the main approaches in the field of HTI today illustrates that it is dominated by a set of more or less well-organized paradigms. Of these, the paradigms related to HCI are much more organized than the others illustrated in this chapter. Of course, the notion of paradigms cannot be considered here in a narrow sense or as a theoretical paradigm only. The paradigms illustrate how the work in the field should be done, and how researchers and designers are organized around the paradigms into more or less rigid groups. Below, other factors that can help structure the field are examined.

From Paradigms to Research Programmes

Traditionally, new paradigms have been developed when important empirical results cannot be adequately explained by the current paradigm (Kuhn 1962). Experiments challenge the concepts of the earlier thinking and test new theories against old ones. For example, Eddington's experiments showed the superiority of relativity theory compared to Newton's thinking.

It is slightly problematic to use the notion of paradigm in the context of HTI and human research (Saariluoma 1997). In the traditional Kuhnian sense, it must be possible to empirically refute a paradigm, and Popper's (1959) and Kuhn's (1962) criterion for scientific work was falsification. However, this criterion makes no sense in the case of HTI approaches. There are no experiments that could prove that, for example, emotional usability should be replaced by some other way of thinking. Emotional usability is a research area rather than a testable theory; therefore, it is not a paradigm in the traditional Kuhnian (Kuhn 1962) sense. Thus, the present generation of apparently new paradigms in the field of HTI is different from what Popper (1959) and Kuhn (1962) described, as HTI paradigms are not created by an inability to explain new empirical results (cf. Saariluoma 1997). Instead, they result from conceptual changes and innovations (Saariluoma 1997), thus they are differentiated and unified by their conceptual base rather than empirical refutations.

The conceptual renewal of technical discourse does not mean that empirical testing is pointless. One can use empirical testing to refute many interaction solutions and theoretical ideas generated within a specific approach. For example, improved screen resolution has made it possible to use a much greater variety of fonts than what was possible in the 1980s, and therefore one could say that the old ways of presenting text information have been refuted. These developments have opened new avenues of research and design; they have not refuted the importance of usability and testing.

There are numerous HTI schools of thought (Carroll 2003; see also Leikas 2009, for an overview), but how do they relate to each other? Some HTI paradigms are conceptually closer to each other than to other paradigms and thus form clusters. For example, ergonomics and human factors belong to the same cluster, but are distant from other paradigms, such as ethical design.

The first cluster is technical paradigms; they are based on technical concepts, which can be typical for WEB design, graphical user interfaces (GUI), or just mechanical machine control. Paradigms in this cluster provide tools for designing artefacts with functionalities and controls that make it possible for people to use technology effectively. For example, technical interaction design tools such as visual C#, UML, ER, and various game engines are widely applied in technical HTI design and research (Booch et al. 1999; Elmasri and Navathe 2011; Lewis and Jacobson 2002). In mechanical engineering, different approaches to control design are equally common. Technical discourses usually do not discuss how people can *best use* technical artefacts, how they *like* to use them, or *what* they use them for; these concepts are left to the designers' intuition within this cluster of paradigms.

The second cluster of paradigms relates to how people can best use technical artefacts. There are numerous approaches to designing controls and other interaction elements in order to make using the artefacts easy, fast, learnable, memorable, and safe, for example (Nielsen 1993). These approaches involve using human research knowledge such as psychology, medicine, and physiology.

Human factors, ergonomics, usability, HCI, and user psychology are examples of paradigms and fields of learning in this cluster; they help solve issues of 'goodness' in HTI design. They are conceptually close to each other as they all prioritize human performance and strive to make it easy to interact with technologies (Carroll 2003; Karwowski 2006; Nielsen 1993).

The third cluster of paradigms relies on emotion-, motivation- and personality-based ways of interaction thinking: making people *like* the technologies. Thus, UX research, affective ergonomics, emotional design, emotional user psychology, and Kansei engineering are unified in their attempts to advance understanding of the emotional basis of interaction (Ciavola et al. 2010; Hassenzahl and Tractinsky 2006; Desmet et al. 2001; Helander and Khalid 2006; Kuniavsky 2003; Leonard and Rayport 1997; Nagamashi 2011; Norman 2004; Rauterberg 2006, 2010). Paradigms such as designing for pleasure, funology, and entertainment computing belong to this cluster (Jordan 2000; Monk et al. 2002). Emotional user psychology is also important here (Saariluoma 2005; Saariluoma and Jokinen 2014).

Many interaction terms such as human- and user-centred design, affective engineering, activity theory, and task analyses also apply to this third cluster of paradigms (Abras et al. 2004; Annett 2000, 2004; Annett and Duncan 1967; Card et al. 1983; Karwowski 2006; Kuutti 1996; Mao et al. 2005; Marti and Bannon 2009; Nardi 1996; Norman and Draper 1986; Saariluoma and Oulasvirta 2010; Stahl 2006, 2010). Ethical design, value-sensitive design, worth-centred development, inclusive design, and design for all can be found in the HTI design literature (Bouma et al. 2009; Charness 2009; Czaja and Nair 2006; Leikas 2009). Much socio-technical research also belongs to this cluster (Trist 1978). All these approaches focus on what people use technologies for.

Metascientist Imre Lakatos (1970) noticed that conceptually similar research projects are often grouped together into *research programmes*. HTI research as a whole can be seen as a collection of paradigms that is organized around different research programmes (Fig. 2.1). Some paradigms are attached to several programmes—for example, affective or cultural ergonomics. Although ergonomics is basically the study of human performance, it also considers important emotional and social themes. Thus, the scope of ergonomics has expanded into the area of other research programmes as they have also been seen as important. In the same way, UX expands on the issues relevant in usability (Nielsen and Norman 2014).

Organizing HTI

Resolution and composition is a renaissance method of thought (Hobbes 1651/1965) that refers to a process in which a topic is cut into pieces and then put together again. In modern terms, one could call it reverse engineering (Chikofsky and Cross 1990). The benefit of reverse engineering is that it helps the engineers understand how a machine or device oper-

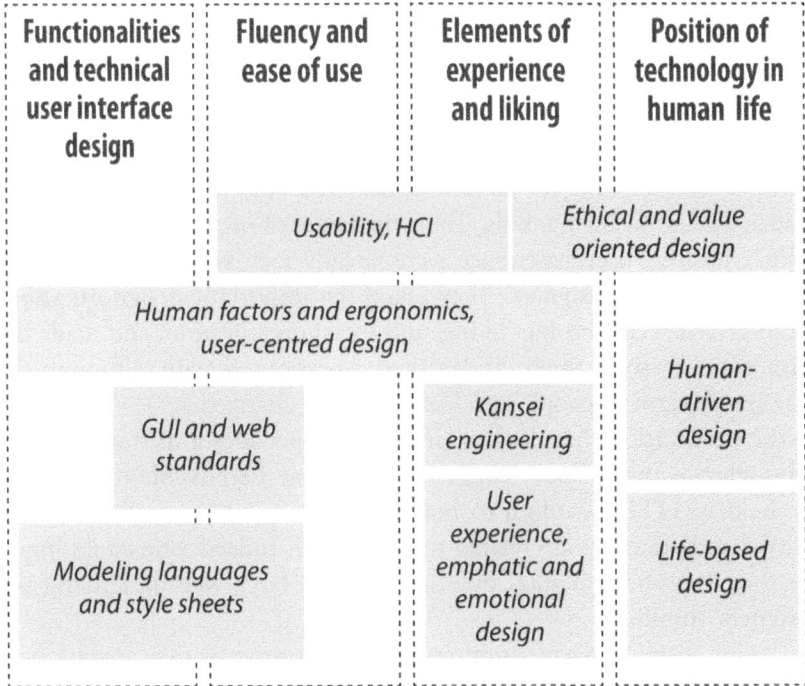

Fig. 2.1 The basic design discourses and respective research programmes

ates, including its components (and their relationships with each other) and their functions. Similarly, categorizing the field of HTI thinking into research programmes and further dividing them into different research questions leads to important insights about how designers' minds work.

Path-breakers in the field of design research have identified practical problems and found ways to solve them. Other designers have shared and modified their problems in their own research, which has led to a set of research projects with more or less similar goals. The field keeps evolving, as new pieces of information are collected in practical empirical research every day.

However, new solutions do not totally revolutionize the problem structure. For instance, the development of touchscreen technology shook up both usability and interaction styles in mobile technology, but designers still had to figure out how to efficiently input knowledge and present feedback, and continue making easy-to-use devices. The technical solutions had changed, but the challenges and problems for designers remained the same.

Stone Age tools were not only practical, they also had additional design features for pleasure and aesthetic reasons. Likewise, in the ancient world as well as today jewels were considered objects of pleasure as well as socially significant symbols. Thus, the issues of organizing life and creating a positive user experience were already tacitly or explicitly in the minds of ancient designers. They asked the important design questions about artefacts concerning 'liking' and social organization, and made the extra effort to solve them. Today these questions remain, although the answers (in term of people's preferences) have changed.

The observation that the fundamental design questions stay the same although the answers vary suggests that it may be possible to structure the field of HTI according to how the basic questions underlying the research programmes are related to each other. Indeed, one could imagine that these fundamental questions form a large web underlying the designers' thinking.

The questions also have another important property: they are task necessary (Saariluoma 1984). This means that people have to solve the problems either consciously or tacitly, but they cannot avoid solving them (Saariluoma 1997). All smartphone designers have to answer numerous questions concerning the attributes of the final product: the form, size, and placement of the touchscreen and the appearance and functions of the icons must all be designed.

Yet is the design of a mobile phone essentially so different from Stone Age designs? In designing a stone axe, people of the time had to consider materials and forms, such as fixing the components together safely. A broken axe was less useful than the original when battling with wolves. Hence, the basic problems of current designs have already been addressed in designing technical artefacts from the start. Of course, there are also a huge number of differences, but the basic issue remains the same: the outcome depends on the questions as much as on the answers.

It is vital to understand that HTI research and design is dominated by the logic of questions, and that basic questions provide a fundamental structure to the field. Questions and answers guide the design and development irrespective of the prevailing design research programmes. HTI introduces a complex set of problems, as the major research programmes differ in what they consider to be the field's most fundamental questions. However, the basic questions of HTI research are complementary. Each of the research programmes adds something essential to the field and thus opens up a new perspective on the problems of HTI.

The idea of using a stable set of questions and their relationships to structure the field of HTI follows what is known about human thinking. The standard psychological definition of thinking is illustrative with respect to this point. As Newell and Simon (1972) put it: 'Thinking is a process which arises when a person has a goal but does not know how to reach it.' Questions, naturally, define these goals, which, together with answering the questions, form the natural structure of human thinking.

Yet the questions are normally not very explicit in design thinking, perhaps because for designers, it may be faster to find design solutions than to address the design questions that should be answered. This kind of tacit structure is common in human knowledge. Grammar, for example, illustrates how people use language and structure sentences, though non-linguists rarely have a clear idea about the structures they follow in building sentences to express their ideas (Chomsky 1957).

The problem of tacit knowledge has historical roots. To understand the hidden structures of our thought, we should explicate tacit knowledge

(Saariluoma 1997). It should be possible to define the tacit dimension of the flow of information, conceptualize it, name it, and incorporate it into the social discourse. Thus, like grammarians, people should be able to give explicit form to the knowledge behind their deeds and thoughts.

Structuring design knowledge by explicating the tacit questions is like a process of writing grammar to support the designers' thinking. The questions express how the designers should pursue better HTI designs. Knowing what the basic problems in a field are, and how they are related to each other, makes it easier to solve the problems. To reach this ultimate goal, it is essential to study the paradigms related to the fundamental problems in order to examine the questions that define the field today.

The Fundamental Four: The Basic Questions and Discourses

The field of HTI research and design has numerous more or less anarchistic paradigms that can be categorized into four major research programmes. These programmes, in turn, are currently determined by four fundamental issues:

• functionalities and technical user interface design;
• fluency and ease of use;
• elements of experience and liking; and
• the position of technology in human actions.

The research and design activity related to these questions can be seen as organized into four discourses, which search for solutions to meet the problems that emerge in practical design and related scientific research. The basic questions can be investigated and answered on the grounds of different types of scientific knowledge. The contents of this book mainly rely on human research (psychology, physiology, action therapy, and sociology), as knowledge of the preconditions and laws of the human mind and actions provides a firm grounding for HTI design. The answers to technical interaction issues, however, mostly rely on machine engineering, programming, and information systems knowledge. The three other fundamental questions (usability, UX, technology in life) can also be examined from the point of view of such applied and human-oriented research traditions as marketing research or economics. Finally, it is also possible to find solutions from important outcomes of innovation research for organizing development processes in industry.

All researchers and designers usually take a position with respect to which research programmes define the key issues in HTI design. Research programmes serve as a basis of social and *design discourses* (Behrent 2013; Ross and Chiasson 2011; Sikka 2011). The word 'discourse' has been actively used among continental philosophers to refer to special ways of thinking and respective special languages (Foucault 1972; Habermas 1973, 1981), which can be about such cultural and political issues as equality of people, freedom, human rights, or power. Design thinking—one of the most important engines for the advancement of human life and society—can also be seen as a discourse with numerous important sub-discourses. The term discourse is particularly illuminating in this context, because design is a social construction process in which numerous people take part in different design discourses and processes. Social media, for instance, entails a variety of blogs, design forums, and discussion groups in which researchers and designers discuss different ideas. Many different kinds of people (and many different kinds of design paradigms) participate in the process of finding unified solutions. Design is a constructive process in the sense that at the beginning, no one knows exactly what the final solution will look like. Understanding of the relevant issues increases during the process. Thus, the outcome is a result of a common thought and communication process (i.e., a design discourse).

In order to gain a deeper understanding of research programmes and their respective design discourses, it is essential to study thought elements, which keep discourses and research programmes together. Each of the four research programmes has its characteristic key questions that researchers and designers must answer. These can, again, be examined with the help of metascience.

The origins of the concept of problem-oriented thinking in investigating scientific thinking can be found in the work of acknowledged metascientist Larry Laudan (1977). He argued that true advancement in science is achieved by improving its capacity to solve problems. Thus, setting and solving problems forms the core of scientific activity. Therefore, to define the structure of HTI paradigms, it is useful to study the basic questions of earlier research programmes.

References

Abras, C., Maloney-Krichmar, D., & Preece, J. (2004). User-centered design. In W. S. Bainbridge (Ed.), *Encyclopedia of human-computer interaction* (pp. 445–456). Thousand Oaks, CA: Sage.

Alexander, C. (1977). *A pattern language: Towns, buildings, construction.* Oxford: Oxford University Press.

Anderson, J. R., Farrell, R., & Sauers, R. (1984). Learning to program Lisp. *Cognitive Science, 8,* 87–129.

Annett, J. (2000). Theoretical and pragmatic influences on task analysis methods. In J. Schraagen, S. Chipman, & V. Shalin (Eds.), *Cognitive task analysis* (pp. 25–40). Mahwah, NJ: Erlbaum.

Annett, J. (2004). Hierarchical task analysis. In D. Diaper & N. Stanton (Eds.), *Handbook of cognitive task design* (pp. 63–82). Hillsdale, NJ: Erlbaum.

Annett, J., & Duncan, K. D. (1967). Task analysis and training design. *Report resumes.* Hull: Hull University.

Aykin, N., Quaet-Faslem, P., & Milewski, A. (2006). Cultural ergonomics. In G. Salvendy (Ed.), *Handbook of human factors and ergonomics* (pp. 3–31). Hoboken, NJ: Wiley.

Baddeley, A. D. (1986). *Working memory.* Cambridge: Cambridge University Press.

Barbour, I. (1980). Paradigms in science and religion. In G. Gutting (Ed.), *Paradigms and revolutions: Appraisals and applications of Thomas Kuhn's philosophy of science* (pp. 223–245). Notre Dame, IN: University of Notre Dame Press.

Behrent, M. C. (2013). Foucault and technology. *History and Technology, 29,* 54–104.

Booch, G., Rumbauch, J., & Jacobson, I. (1999). *The unified modelling language.* Reading, MA: Addison-Wesley.

Bouma, H., Fozard, J. L., & van Bronswijk, J. E. M. H. (2009). Gerontechnology as a field of endeavour. *Gerontechnology, 8,* 68–75.

Bowen, W. R. (2009). *Engineering ethics: Outline of an aspirational approach.* London: Springer.

Bridger, R. S. (2009). *Introduction to ergonomics.* Boca Raton, FL: CRC Press.

Broadbent, D. (1958). *Perception and communication.* London: Pergamon Press.

Budd, T. (1991). *Object-oriented programming.* Reading, MA: Addison-Wesley.

Card, S., Moran, T., & Newell, A. (1983). *The psychology of human-computer interaction.* Hillsdale, NJ: Erlbaum.

Carroll, J. M. (Ed.). (2003). *HCI models, theories, and frameworks: Toward a multidisciplinary science.* San Francisco, CA: Morgan Kaufmann.

Charness, N. (2009). Ergonomics and aging: The role of interactions. In I. Graafmans, V. Taipale, & N. Charness (Eds.), *Gerontechnology: Sustainable investment in future* (pp. 62–73). Amsterdam: IOS Press.

Chikofsky, E. J., & Cross, J. H. (1990). Reverse engineering and design recovery: A taxonomy. *Software, IEEE, 7*, 13–17.

Chomsky, N. (1957). *Syntactic structures.* The Hague: Mounton.

Ciavola, B., Ning, Y., & Gershenson, J. K. (2010). Empathic design for early-stage problem identification. In *American Society of Mechanical Engineers 2010 International Design Engineering Technical Conferences and Computers and Information in Engineering Conference* (pp. 267–276).

Cockton, G. (1987). Interaction ergonomics, control and separation: Open problems in user interface management. *Information and Software Technology, 29*, 176–191.

Cockton, G. (2004). Value-centred HCI. In *Proceedings of the Third Nordic Conference on Human-Computer Interaction* (pp. 149–160).

Coursaris, C. K., & Bontis, N. (2012). A metareview of HCI literature: Citation impact and research productivity rankings. In *SIGHCI 2012 Proceedings Paper 9*.

Covan, N. (2000). The magical number 4 in short-term memory: A reconsideration of mental storage capacity. *Behavioural and Brain Sciences, 24*, 87–185.

Cross, N. (1982). Designerly ways of knowing. *Design Studies, 3*, 221–227.

Cross, N. (2001). Designerly ways of knowing: Design discipline versus design science. *Design Issues, 17*, 49–55.

Cross, N. (2004). Expertise in design: An overview. *Design Studies, 2*, 427–441.

Czaja, S. J., & Nair, S. N. (2006). Human factors engineering and systems design. In G. Salvendy (Ed.), *Handbook of human factors and ergonomics* (pp. 32–49). Hoboken, NJ: Wiley.

Deacon, T. (1997). *The symbolic species: The co-evolution of language and the human brain.* London: Penguin Press.

Desmet, P., Overbeeke, K., & Tax, S. (2001). Designing products with added emotional value: Development and application of an approach for research through design. *The Design Journal, 4,* 32–47.

Di Stasi, L. L., Antolí, A., & Cañas, J. J. (2013). Evaluating mental workload while interacting with computer-generated artificial environments. *Entertainment Computing, 4,* 63–69.

Dijkstra, E. (1972). Notes on structured programming. In O. Dahl, E. Dijkstra, & C. Hoare (Eds.), *Structured programming.* London: Academic Press.

Dix, A., Findlay, J., Abowd, G., & Beale, R. (1993). *Human-computer interaction.* New York: Prentice-Hall.

Dym, C. L., & Brown, D. C. (2012). *Engineering design: Representation and reasoning.* New York: Cambridge University Press.

Elmasri, R., & Navathe, S. (2011). *Database systems.* Boston, MA: Pearson Education.

Feyerabend, P. (1975). *Against method.* London: Verso.

Forlizzi, J., & Battarbee, K. (2004). Understanding experience in interactive systems. In *Proceedings of the 5th Conference on Designing Interactive Systems: Process, Practices, Methods, and Techniques (DIS 2004)* (pp. 261–268).

Foucault, M. (1972). *The archaeology of knowledge and the discourse on language.* New York: Pantheon Books.

Galiz, W. O. (2002). *The essential guide to user interface design.* New York: Wiley.

Gopher, D., & Donchin, E. (1986). Workload: An examination of the concept. In K. R. Boff, L. Kaufman, & J. P. Thomas (Eds.), *Handbook of perception and human performance: Cognitive processes and performance* (pp. 1–46). Hoboken, NJ: Wiley.

Habermas, J. (1973). *Erkentniss und interesse* [Knowledge and interests]. Frankfurth am Main: Surkamp.

Habermas, J. (1981). *Theorie des kommunikativen Handelns* [Theory of communicative behavior] (Vols. 1–2). Frankfurt am Main: Suhrkamp.

Hanson, N. R. (1958). *Patterns of discovery.* Cambridge: Cambridge University Press.

Hassenzahl, M. (2011). *Experience design.* San Rafael, CA: Morgan & Claypool.

Hassenzahl, M., & Tractinsky, N. (2006). User experience—A research agenda. *Behaviour and Information Technology, 25,* 91–97.

Helander, M., & Khalid, H. M. (2006). Affective and pleasurable design. In G. Salvendy (Ed.), *Handbook of human factors and ergonomics* (pp. 543–572). Hoboken, NJ: Wiley.

Hobbes, T. (1651/1950). *Leviathan*. New York: EP Dutton.

International Organization for Standardization. (1998). *ISO 9241-11: Ergonomic Requirements for Office Work with Visual Display Terminals (VDTs): Part 11: Guidance on Usability.*

Jordan, P. W. (2000). *Designing pleasurable products: An introduction to the new human factors*. Boca Raton, FL: CRC Press.

Kaptelinin, V. (1996). Activity theory: Implications for human-computer interaction. In B. A. Nardi (Ed.), *Context and consciousness: Activity theory and human-computer interaction* (pp. 103–116). Cambridge, MA: MIT-Press.

Karwowski, W. (2006). The discipline of ergonomics and human factors. In G. Salvendy (Ed.), *Handbook of human factors and ergonomics* (pp. 3–31). Hoboken, NJ: Wiley.

Kieras, D. E., & Meyer, D. E. (1997). An overview of the EPIC architecture for cognition and performance with application to human-computer interaction. *Human-Computer Interaction, 12*, 391–438.

Koivisto, K. (2011). Kaj Frank and the art of glass. In H. Matiskainen (Ed.), *The art of glass—Kaj Frank 100 years* (pp. 8–61). Saarijärvi: Design museo.

Kuhn, T. (1962). *The structure of scientific revolutions*. Chicago: University of Chicago Press.

Kuniavsky, M. (2003). *Observing the user experience: A practitioner's guide to user research*. San Mateo, CA: Morgan Kaufmann.

Kuutti, K. (1996). Activity theory as a potential framework for human–computer interaction research. In B. A. Nardi (Ed.), *Context and consciousness: Activity theory and human–computer interaction* (pp. 17–44). Cambridge, MA: MIT Press.

Lakatos, I. M. (1970). Falsification and the methodology of research programmes. In I. Lakatos & A. Musgrave (Eds.), *Criticism and the growth of knowledge*. Cambridge: Cambridge University Press.

Laudan, L. (1977). *Progress and its problems: Towards a theory of scientific growth*. London: Routledge and Kegan Paul.

Leikas, J. (2009). *Life-based design—A holistic approach to designing human-technology interaction*. Helsinki: Edita Prima Oy.

Leonard, D., & Rayport, J. F. (1997). Spark innovation through empathic design. *Harvard Business Review, 75*, 102–115.

Lewis, M., & Jacobson, J. (2002). Game engines. *Communications of the ACM*, *45*, 27–31.

Mao, J., Vredenburg, K., Smith, P. W., & Carey, T. (2005). The state of user-centered design practice. *Communications of the ACM*, *4*, 105–109.

Mattelmäki, T., & Battarbee, K. (2002). Empathy probes. In: T. Binder, J. Gregory and I. Wagner (Eds.), Proceedings of the 7th Biennial Participatory Design Conference 2002, June 23 - June 25, 2002, Malmø, Sweden. (pp. 266–271).

Markopoulos, P., & Bekker, M. (2003). Interaction design and children. *Interacting with Computers*, *15*, 141–149.

Marti, P., & Bannon, L. J. (2009). Exploring user-centred design in practice: Some caveats. *Knowledge, Technology and Policy*, *22*, 7–15.

McCarthy, J., & Wright, P. (2004). Technology as experience. *Interactions*, *11*, 42–43.

Miller, G. A. (1956). The magical number seven, plus or minus two: Some limits on our capacity for processing information. *Psychological Review*, *63*, 81–97.

Monk, A. F. (2002) Fun, communication and dependability: Extending the concept of usability. In Faulkner, X, Finlay, J, and Detienne, F. (Eds)16th British-Human-Computer-Interact-Group Annual Conference/European-Usability-Professionals-Association London, England Sept. 02-06, 2002, p. 3–14.

Nagamashi, M. (2011). Kansei/affective engineering and history of Kansei/affective engineering in the world. In M. Nagamashi (Ed.), *Kansei/affective engineering* (pp. 1–30). Boca Raton, FL: CRC Press.

Nardi, B. A. (1996). *Context and consciousness: Activity theory and human-computer interaction*. Cambridge, MA: MIT Press.

Newell, A., & Simon, H. A. (1972). *Human problem solving*. Engelwood Cliffs, NJ: Prentice-Hall.

Nielsen, J. (1993). *Usability engineering*. San Diego, CA: Academic Press.

Nielsen, J., & Norman, D. (2014). Definition of user experience. Retrieved January 24, 2015, from http://www.nngroup.com/articles/definition-user-experience/

Niezen, G. (2012). *Ontologies for interaction*. Eindhoven: Eindhoven University Press.

Norman, D. (2004). *Emotional design: Why we love (or hate) everyday things*. New York: Basic Books.

Norman, D. A., & Draper, S. W. (Eds.). (1986). *User centered system design; New perspectives on human-computer interaction*. Hillsdale, NJ: Erlbaum.

Norman, D., Miller, J., & Henderson, A. (1995). What you see, some of what's in the future, and how we go about doing it: HI at Apple Computer. In *Conference Companion on Human Factors in Computing Systems* (p. 155). ACM.

Popper, K. R. (1959). *The logic of scientific discovery*. London: Hutchinson.

Preece, J., Rogers, Y., Sharp, H., Benyon, D., Holland, S., & Carey, T. (1994). *Human-computer interaction*. Harlow: Addison-Wesley.

Preece, J., Rogers, Y., & Sharp, H. (2004). *Interaction design*. New York: Wiley.

Rauterberg, M. (2004). Positive effects of entertainment technology on human behaviour. In: R. Jacquart (Ed): Building the Information Society (pp. 51-58). Kluwer Academic press.

Rauterberg, M. (2006). HCI as an engineering discipline: To be or not to be? *African Journal of Information and Communication Technology, 2,* 163–183.

Rauterberg M. (2010). Emotions: The voice of the unconscious. In: H.S. Yang, R. Malaka, J. Hoshino, J.H. Han (eds.) Entertainment Computing - ICEC 2010 (Lecture Notes in Computer Science, vol. 6243, pp. 205–215), (c) IFIP International Federation for Information Processing, Heidelberg: Springer.

Ross, A., & Chiasson, M. (2011). Habermas and information systems research: New directions. *Information and Organization, 2,* 123–141.

Saariluoma, P. (1984). Coding problem spaces in chess. In *Commentationes Scientiarum Socialium* (Vol. 23). Turku: Societas Scientiarum Fennica.

Saariluoma, P. (1997). *Foundational analysis: Presuppositions in experimental psychology*. London: Routledge.

Saariluoma, P. (2005). Explanatory frameworks for interaction design. In A. Pirhonen, H. Isomäki, C. Roast, & P. Saariluoma (Eds.), *Future interaction design* (pp. 67–83). London: Springer.

Saariluoma, P., & Jokinen, J. P. (2014). Emotional dimensions of user experience: A user psychological analysis. *International Journal of Human-Computer Interaction, 30,* 303–320.

Saariluoma, P., & Oulasvirta, A. (2010). User psychology: Re-assessing the boundaries of a discipline. *Psychology, 1,* 317–328.

Saarinen, E., & Saarinen, A. B. (1962). *Eero Saarinen on his work*. New Haven, CT: Yale University Press.

Schmidt, R. A., & Lee, T. D. (2011). *Motor control and learning: A behavioral emphasis*. Champaign, IL: Human Kinetics.

Seligman, M. E. P., & Csikszentmihalyi, M. (2000). Positive psychology—An introduction. *American Psychologist, 55,* 5–14.

Shneiderman, B., & Maes, P. (1997). Direct manipulation vs. interface agents. *Interactions, 4,* 42–61.

Shneiderman, B., & Plaisant, C. (2005). *Designing user interfaces*. Boston, MA: Pearson.

Sikka, T. (2011). Technology, communication, and society: From Heidegger and Habermas to Feenberg. *The Review of Communication, 11*, 93–106.

Simon, H. A. (1969). *The sciences of artificial*. Cambridge, MA: MIT Press.

Stahl, B. C. (2006). Emancipation in cross-cultural IS research: The fine line between relativism and dictatorship of intellectual. *Ethics and Information Technology, 8*, 97–108.

Stahl, B. C. (2010). 6. Social issues in computer ethics. In L. Floridi (Ed.), *The Cambridge handbook of information and computer ethics* (pp. 101–115). Cambridge: Cambridge University Press.

Trist, E. L. (1978). *On socio-technical systems. Sociotechnical systems: A sourcebook.* San Diego, CA: University Associates.

Väänänen-Vainio-Mattila, K., Väätäjä, H., & Vainio, T. (2009). Opportunities and challenges of designing the service user eXperience (SUX) in web 2.0. In P. Saariluoma & H. Isomäki (Eds.), *Future interaction design II* (pp. 117–139). Berlin: Springer.

van Schomberg, R. (2013). A vision of responsible research and innovation. In R. Owen, M. Heintz, & J. Bessant (Eds.), *Responsible innovation* (pp. 51–74). Oxford: Wiley.

Vygotsky, L. S. (1980). *Mind in society: The development of higher psychological processes*. Cambridge, MA: Harvard University Press.

Weinberg, G. M. (1971). *The psychology of computer programming*. New York: Van Nostrand Reinhold.

Whitford, F., & Ter-Sarkissian, C. (1984). *Bauhaus*. London: Thames and Hudson.

Wickens, C., & Holands, J. G. (2000). *Engineering psychology and human performance*. Upper Saddle River, NJ: Prentice-Hall.

3

The Logic of User Interface Design

Technical artefacts exist so that people can use them to make something happen. Their capacity to do so depends on the functions and functionalities of the technology, which requires users. Technologies thus have to give users the ability to control them, and the designer's role is to create the actions and work processes for which the artefacts are intended. This basic HTI pursuit is called user interface design. It applies technical interaction concepts to solve design problems. This chapter presents the overall principles and goals for the user interface design of any technical artefact.

Technical artefacts are physical systems that only follow the laws of physics; they are not able to set sense-making goals for their operation. Only people can. All possible goals (end states) are equally 'valuable' for artefacts. Therefore from the point of view of the laws of physics, it is irrelevant whether an aeroplane lands successfully or crashes. Yet from a human point of view, a successful landing is desirable and a crash is a catastrophe. Sensible behaviours of any technical artefacts are thus a sub-set of all their possible behaviour alternatives. Since artefacts as physical systems cannot distinguish between successful and non-successful behaviours, it is essential that people control them.

© The Editor(s) (if applicable) and The Author(s) 2016
P. Saariluoma et al., *Designing for Life*,
DOI 10.1057/978-1-137-53047-9_3

It is essential to differentiate between human intentions (the goals people pursue) and the expected goal state of a machine. The latter is the state of an artefact that makes it possible to support human goal-driving or intentional actions. For example, people use computers to make transactions in banks. Transferring money to pay an invoice is an intentional human action, and its goal is to pay the invoice. However, the ability to transfer money from one's own bank account to that of the invoicing company requires using the bank's software system.

The two actions—making the transaction and using the bank's software—are not identical, as they have different goals. Transferring money is the goal of human action, and being able to carry out the operations required by the software is another action, namely the aim to reach the goal state of the technical artefact. There are thus two different, albeit intertwined, questions: first, what are human actions, and second, how should technical artefacts be controlled to realize them.

The user interface is the necessary part of a technical artefact (e.g., a computer or mechanical device) that people use to control its behaviour. Its main function is to connect a human being and the behaviour of the artefact. The problem of how user interfaces are controlled is task necessary.

Technical artefacts need information to direct their behaviour towards a sense-making outcome. Users feed in information in order to control the behaviour of the artefacts. In this respect, scissors have the same conceptual structure as computers. They have a handle and finger holes as input modes, blades for operating the target, and a pivot point with a screw for 'internally' controlling the operations. When using scissors, people control how the cutting proceeds, and at the same time use input modes to create the desired outcome.[1] Thus information input by people is used to direct the process of technical artefacts towards the expected goal states.

Technical artefacts have an unlimited number of *possible states*, some of which are *expected goal states*. They are the result of a number of *technical*

---

[1] To be exact, scissors are analogous machines that can have an infinite number of possible input and output states. However, they only reach a finite number of possible states before they are destroyed.

processes. A ship transporting goods from one continent to another is in its expected goal state when it has reached the harbour to unload its cargo. However, the goal is reachable only through the process of sailing off-shore. The process and the goal form a unified whole, which is expressed in the final goal state. Using technology requires defining what users are expected to do during the process. The first step in solving how-to problems is to define the goal and the respective goal-directed processes—the respective tasks and sub-tasks in using technology.

Designers investigate people's action goals in order to create good technologies. They define the role and function of an artefact and adjust its operations to human actions, which are organized into sub-actions that seek to achieve a goal or intention: the *user need*. The function of technology is to help people pursue their action goals. Thus the structure of the action or *task* defines what kind of technical artefacts can support the action.

A concrete and familiar example of user need and the respective user interface is the case of a lift. In a block of flats, the elevator user's *action goal and intention* is to get to another floor. Using the staircase can achieve this goal, but this is a slow and difficult option, especially if one is headed to the top floor. A lift is a suitable technical solution to support movement between the floors. Moving between floors is thus a human action, which has a goal and a respective *user need*. The action itself entails a number of sub-actions such as moving to the lift and entering it. Controlling the behaviour of the lift, or simply *using* it, is one of the sub-actions of moving with it to another floor.

Technical artefacts have technical properties that enable people to reach their goals. Lifts, for example, have technical *components and capacities*, such as power machines and control and safety systems, to move certain weights from one level to another. They also have user interfaces, which allow users to control their behaviour. Technology design discourse as a whole is a combination of these two aspects of the design process. This chapter focuses on user interfaces.

The first question in HTI design concerns *how the behaviour (functionalities) of a technical artefact can be controlled*. How should the artefact be manoeuvred so that it can reach its *expected state* or *carry out the expected processes*? During a human action, an expected state can refer to a process that makes it possible for the user to reach her goal. In this sense, the expected state of a sailing boat can be as much about sailing on the sea as reaching a destination.

How to control the behaviour of a technical artefact is a fundamental problem in HTI design. No technical artefact can exist without providing its users the methods to use it. Despite its apparent simplicity, the problem of *how to control* (a technical artefact) has its notable dimensions. First, the behaviour of the artefact must be logically linked to the human action in question. In the case of the lift, the technical capacity to move from one floor to another is the artefact behaviour that makes it possible to support people's movement in a block of flats. Second, it is essential to link the behaviour of the artefact to users' actions via a user interface. In the case of a lift, this often refers to the set of control buttons referring to which floor the lift should stop on. However, the latter presupposes design knowledge of how people use the artefact.

Functions and Events

Task analysis, the schematic structure it gives to tasks and the actions of using a technology form the basis of technology design. The role of a technical artefact in human actions is normally called its *function*, and its respective capacities are known as *functionalities* (Gero 1990; Gero and Kannengiesser 2004; Ulrich and Eppinger 2011). One could say that functionalities detail the uses of the artefact. As a property of a technical artefact or system, they form the first step in defining technical interaction (Chandrasekaran 1990; Gero 1990; Gero and Kannengiesser 2014). They express the performance capacities of the designed artefact (Gero 1990). In the case of a window, for example, the notion of functionality refers to such properties as providing daylight, making ventilation possible, preventing heating loss, and eliminating noise.

The concept of function in HTI discourse explains what technology can offer to people. Sometimes researchers also use the notion of affordance, the origins of which can be found in Gibson's (1979) psychology of perception (Gaver 1991). Functionality defines the effects that operating an artefact can have on the environment. The goodness or purposefulness of an artefact is also often expressed in terms of its functionality. For example, if a window does not insulate well against frost, it has poor functionality.

The concept of functionality is important in designing different subsystems. For example, in a graphic user interface (GUI), many elements such as menus and dialogue boxes have the function that they can be used to select possible courses of action (Cooper et al. 2007; Griggs 1995). For example, in a mobile phone, the icon of a camera indicates that one can switch the phone to camera mode and take photographs. Similarly, the icon of a compass indicates opening a navigation service. Thus, the main screen may have numerous icons and menus that all have their own functionalities and open up different ways of operating with a given technology.

Functionalities in design discourse define the design *requirements*— what an artefact should be able to do. An interaction design requirement is a goal asset in the design process. Requirements express what a new product should be like and what kinds of needs it is supposed to meet. If the ability to take photographs with a tablet is a user need, it becomes a requirement for designers and a functionality of the product.

On a basic how-to-control level, functionalities define the goals for immediate interaction when using a given technology. The user inputs a command by, for example, pushing a button on a device, or feed-specific information into the system, which in turn steers the behaviour of the artefact. The act of controlling an artefact can be called an *event* or a *control event* in a physical or information system (Datye 2012; Memon et al. 2001; Myers et al. 2000). An event is a significant change in the state of an artefact towards the expected goal state. Event-driven thinking is mostly used in software research and in designing mechanical artefacts, but it can also be used, for example, in service business.

An event thus involves an individual human act of controlling the state and behaviour of an artefact, for example by pressing a start button in an engine or turning a control knob. In technical discourse, events and event handlers are often discussed; these are programs that transform a technical artefact into an expected goal state (Memon et al. 2001). Every piece of information that a user provides a technical artefact with is an event. The concept of an event could also refer to opening up a dialogue box, entering information, resizing a window, or moving it with a mouse. In other cases, the physical interaction of an event could be pressing a lever or turning a steering wheel. In brief, an event or interaction event represents the moment when a user has an effect on the system.

An event has two main components: a triggering action and a respective response in a device or system. The former gives the user the ability to choose the next expected state of the artefact. The latter is a process that moves the artefact or system from a given initial state to the expected goal state. Thus, using a technology means controlling *event flows* from the initial state of a technical artefact to an expected goal state (Memon et al. 2001). The user's role is to choose between possible courses of action.

A series of events forms control histories, which event handlers coordinate. The operation of artefacts proceeds from one state to another, and eventually reaches the final goal state. One can also call this kind of sequence of interlinked events as *event flows, event histories,* or *event paths* (Fig. 3.1). The structure of these interaction processes forms event trees, which illustrate alternative paths for reaching different goal states of the technical artefact. In technical design, all events and the flows needed for the behaviour of the artefact are designed, analysed, and tested in practice to prove the logic of events (Memon et al. 2001; Yuan et al. 2011). On web pages, the event tree can be called a site map (Lin et al. 2000; Newman and Landay 2000). Site maps are organized lists of web pages

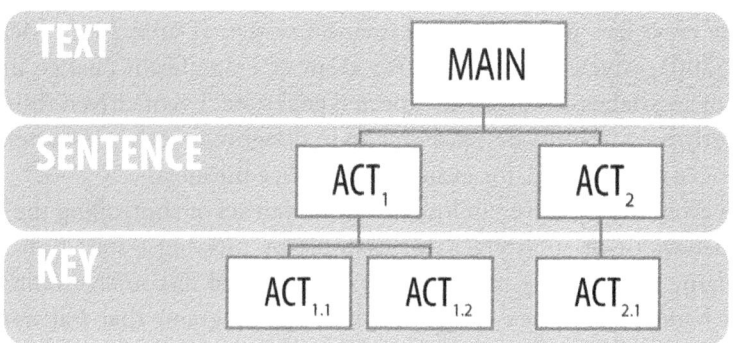

Fig. 3.1 Event paths in an event tree—an example of text processing

that are linked together in a single site. They aid navigation and make it possible for users and search engines to find pages with related content.

An event flow is a sense-making concept in designing user interfaces and respective processes of technical artefacts. In order to slow down the speed of a car, the driver has to choose a lower gear. Doing this requires pressing the clutch to shift into another gear. This kind of common control operation is similar to interacting with GUIs. Events must be defined so that the user can control the technical artefact in a sense-making manner.

Functionalities, requirements, events, and event flows are intimately related to technical concepts. They define the potential processes of a technical artefact and how users can control machine processes or choose between alternative event paths, so that the artefact can do what it is supposed to do. Their design is based on technical theory languages. The critical design question is to define the best structure for event flows from an initial state to a goal state in realizing a particular function (Rauterberg 1996; Ulich et al. 1991).

Event flows can be presented as graphs. In these graphs, nodes represent the input actions of the user and arrows the transformation of the technical artefact from one state to another. On this level the interaction process can be understood as a state-action graph; such graphs have often been used to describe human mental representations (Newell and Simon 1972). This is an equally valid conceptualization for both mechanical and information systems. Thus, riding a bike can be presented as a graph of turning left and right and using hand or pedal brakes. These actions are analogical—that is, they are not really discrete states but can be approximated by finite state graphs just as, for example, voices can be digitalized. In information systems, finite state structures are natural.

From the user's point of view, information systems and other computational technologies are finite by nature. Moreover, the nodes in an event tree entail two types of information, indicating the two modes that users have. First, users give control and steering information to the system by 'telling' the system what the next goal is (mostly by using dialogue boxes or menus). Second, this type of information is relevant for actual tasks. For example, editing a text presupposes writing the text and using the different control facilities of a text-editing program. No interaction process can be defined without investigating the concepts of functionalities and event flows. Even when designing a simple tool such as a saw, these notions are necessary. One function of a saw is to cut down trees. Taking a saw in hand to carry this out entails an event flow. However, *necessary concepts* are not necessarily *sufficient concepts*. To make sense of what the right functionalities and events would be, it is essential to consider the nature of the intended tasks and their related user requirements. Here, task and work analyses can illustrate the functions and goals of events in event trees (Annett 2000; Crandall et al. 2006; Rasmussen et al. 1994; Stanton 2006; Vicente 1999).

Understanding the User's Tasks

The first step towards defining how technology should perform is to define precisely how a new technical artefact should function in relation to specific human actions (i.e., what the users do and how they use the given artefact to reach their goals). This process can be carried out using a *task analysis* (Annett 2004; Annett and Duncan 1967). Designers use task analyses to define the goals and structures of the users' actions and sub-actions in order to develop a technology concept and define the functions of the technical artefact. This information is needed to construct sense-making technical solutions.

Using (operating) a technical artefact is one element of the whole action or task. For example, a radiographer has to carry out certain actions when taking an X-ray, including positioning the instrument, taking the X-ray, downloading and examining it on a computer, and interpreting it. Thus using the instrument forms one part of the whole operation.

The task analysis process was created by Taylor (1911) to make work processes more rational. Since the early 1970s it has attracted increasing attention in technology design, and numerous sophisticated variations of the process have been developed; it has also been used in larger settings for work analyses (Annett 2000, 2004; Annett and Duncan 2000; Diaper 2004; Hollnagel 2006; Rasmussen et al. 1994; Stanton 2006; Vicente 1999).[2]

---

[2] We do not discuss here the differences between task and work analysis, though the difference is essential in the context of work processes. This section focuses on technical interaction, not on the way work groups are organized.

Task analysis is a suitable procedure for examining HTI issues in all fields of technology. Lately, the method has received considerable attention in the area of designing human–computer interaction (HCI) and information systems (Card et al. 1983; Karwowski 2006; Nielsen 1993; Nickerson and Landauer 1997; Rosson and Carroll 2002; Woods and Roth 1988). It is also still widely used in traditional mechanical engineering (Czaja and Nair 2006; Sanders and McCormick 1993). The usage of cars, trains, and cranes, as well as white goods and domestic appliances can be examined equally well with the help of task analysis.

The main goal of task analysis is to rationally stipulate the task (the action in question) and define how a particular technical artefact will be controlled. The ultimate goal is to design how people and technical artefacts can be coordinated in action. Reaching this goal requires defining the following in advance: the series of actions users have at their disposal, the goals of the actions, and the components or sub-actions typical of this action. For example, when developing patient monitoring systems it is important to define what a target group—for example, medical staff in an operating room—does, and what kind of sub-actions they consider relevant during an operation in order to be able to understand the roles of the supporting technology. Similarly, by understanding how a team of nurses and doctors operates in primary health care it would be possible to outline roles for supporting information systems in a primary care clinic and to get the best output from both the people and the technology (Hysong et al. 2011).

Task analysis breaks the higher-level main task into rational sub-tasks (Annett 2004; Annett and Duncan 1967) using empirical information collected by, for example, observation, interviews, focus groups, and various co-design activities (Kjeldskov and Paay 2012; Markopoulos and Bekker 2003; Morgan 1996; Smith 2003). This information can be used to design improvements for actions in general as well as particular work processes (Annett 2000; Hollnagel 2006; Rasmussen et al. 1994; Stanton 2006; Vicente 1999). This process generates a hierarchical representation of what people do and what kind of actions they perform in a certain context to achieve a particular goal (Annett 2004; Annett and Duncan 1967).

A familiar example of investigating an elementary interaction process is typing: the process of writing a text is the high-level (main) action.

It can be divided into different sub-actions that are executed when typing a sentence—such as *insert, replace text, save, delete, shift line,* and *access text from a file* (Annett 2000; Card et al. 1983; Hollnagel 2006). These sub-actions can be analysed in terms of the keystrokes used (e.g., pushing a line feed or using the delete button). In this way it is possible to model the process of typing down to the keystroke level, or even down to the level of the typists' muscle control. This method can be used to represent the hierarchical actions and sub-actions in any human task, from typing to steering a ship (Annett 2004). Yet task analysis can also be used to conceptualize tasks in many other kinds of conceptual systems, such as scenarios (Go and Carroll 2004) or computational models (Card et al. 1983; Kieras 1997). Although their perspectives may be different, all these models of task analysis work to describe tasks as combinations of units. They differ in their characterization of the elements and the way they are unified.

When defining the structure of human actions for a specific task, it is equally important to consider people's goals and the purposes of their actions. Human actions are always intentional and goal directed (Brentano 1874/1955; Dennett 1991; Husserl 1901–1902; Searle 1992). People have reasons for what they do, which are usually explained by the goals they wish to reach (von Wright 1971; Taylor 1964). People do not use TV remote controls to play with or control technology. Instead, they use them to access news or movies on TV. Each action has its goal, for example watching a movie, and this action generates sub-actions to reach this goal, in this case pushing the buttons of a remote control.

Task analysis explicates tasks as a function of sub-tasks: the main action is realized to reach the goal of the main task. However, reaching a goal presupposes carrying out all the necessary sub-tasks and sub-actions. The goals of the sub-tasks are defined by their purpose in helping to achieve the main task. In this sense, sub-tasks and their related actions become meaningful only as a part of the main task. If a father wants to buy a book as a present for his child, he may get in a car and drive to a bookshop, buy a book and drive back home. The purpose of buying a book may be to make his child happy or to encourage his mental development. Going to a bookshop requires taking a car. Buying a book presupposes paying for it and, quite likely, using a credit card. Thus the sub-tasks and sub-actions are ways to reach the goal of the main task, and their purpose is defined as part of the main task.

The core of task analyses is to illustrate the hierarchical action and sub-action sequences or histories in a detailed task hierarchy. It requires defining the functions of the actions in using the artefact, and their mutual relations and sub-actions as a whole, as this knowledge is needed to outline requirements and solutions for technologies supporting the actions. The functional structure of a task thus provides designers with information about design requirements, which can be used to explain why individual design solutions make sense and what their optimal form should be.

A proper task analysis helps designers decide which kinds of technical solutions and user requirements are needed to accomplish the chosen task: that is, the kind of functional roles the technology could have and the kind of opportunities the user should be given to control the operation of the artefact. The performance capabilities needed for technology depend on the nature of the task and the human intentions in carrying it out. Task analysis provides this information about what the new technology is supposed to do.

Intentional and Hidden Interaction Elements

Modern HTI can be constructed in two different ways: intentional or tacit. Often the interaction is supposed to be explicit and conscious. In this type of interaction, users need to understand the behaviour of an artefact and how their intentional involvement affects this behaviour. Users' control of actions and the operations of the machine must be clearly linked to each other and understood by the users. Users must be given all the necessary information and all the means to control the process in all possible scenarios. This kind of interaction can be called *intentional*, as it is based on the conscious intentions of the users. Users know what they would like to achieve—what the goal is—and how they should use technical artefacts to do so.

Direct manipulation schemes provide a good example of intentional interaction (Shneiderman 1983; Shneiderman and Maes 1997). In direct manipulation, users are given every means to directly influence the way technical artefacts operate, and to be aware of how operations precede. Users are allowed to use visible or directly sensible objects that are directly linked to the operations of the artefacts. When direct manipulation is allowed, computer users can, for example, drag and drop items or change their sizes at will. Today many technologies have internal and tacit user-interaction solutions, which are used to help control the behaviour of a technical artefact. Cars, for example, may have numerous tacit control systems of which users are barely aware. These systems operate ubiquitously in the background and make driving easier, although the driver has no direct contact with them (or a need to know what is happening in the system).

Also in some devices that resemble a bike, such as balance electric scooters, it is possible to automate a balance-keeping function with a smart computer program. This kind of automation represents a common interaction design technique. For example, when a smartphone is picked up, the device adapts the display so the user can easily read the information on the screen. Repositioning the phone thus indirectly activates automatic operations that present the content on the display in an optimal manner. This type of machine operation, which directs the machine towards an anticipated and optimal state, can be called *internal* or *machine interaction operations*.

Another example of tacit functions is related to navigation on a screen. Main menus usually provide limited information and hide the rest until selected by the user. The users must thus have an idea about what is hidden and what kind of information can be expected behind the selections. In the design, it is important to perceive the correct order for this kind of invisible information.

Different kinds of memory support systems, such as balloon tips, can help users navigate such menus. However, current user interface solutions often do not provide particularly efficient solutions due to technical limitations. Limits on the space for GUI or the size of touchscreens, for example, makes it necessary to hide certain functionalities and commands from the primary display. Therefore finding the commands must be so intuitive, and the information concerning them so purposeful, that people can reach their task goals using the given controls. This is possible only when user tasks, and the structure of particular actions by means of tasks analyses, are intuitively perceived.

Arguably, the most important form of tacit interaction today is ubiquitous computing (Weiser 1993). In such variants as pervasive computing or computing everywhere, the traditional command- or GUI-based interfaces are replaced by interfaces that automatically register the behaviour of the user and respond accordingly. This kind of interaction can support human actions without the active involvement of users.

Ubiquitous computing is an important paradigm in HTI, as it enables people to use technologies without special training or skills. These technologies aid people in their actions unnoticed and make the users' goals easier to reach. Ubiquitous computing can be used, for instance, in gerontechnology to monitor the health of older people or to adapt the living environment according to changes in human behaviour in order to prevent accidents (Bouma et al. 2009; Charness 2009; Czaja and Nair 2006; Leikas 2009).

Intentional and tacit interaction always presupposes the analysis of tasks. Most of the current application programs have numerous tacit elements, though their overall use is intentional. The crucial questions in design are associated with dividing tasks between intentional and tacit processes, as well as recognizing the relevant characteristics of human factors that can be used to launch tacit technical processes.

User Interfaces: The Means of Control

User interfaces are systems of signs that have the same structure as event trees. They also have tacit and intentional components, and they are connected to the actions people carry out when they use the interfaces and technologies in question. User interfaces are always created by synthesizing different types of man-machine discourses.

The core of an interface concerns selecting expected tasks for a given technology. Before people can benefit from any technology, it must be used; thus the technical artefact must have a user interface. Standard lifts, for example, have buttons to push to access different floors. Users select the preferred floor and push the corresponding button on an elevator interface. Developing a lift—or any technical artefact—without a means of controlling it would not make sense.

The design of a user interface and the associated controls of the artefact's behaviour form two of the basic problems in HTI design. Ultimately, user interfaces give people the ability to control technical artefacts. This is based on the interaction between control equipment and human actions in different types of practical situations, and requires an understanding of the expected behaviour and states of the artefact in relation to different types of human actions.

The nature of human actions justifies the use of a certain technical artefact. The control systems must be based on knowledge of the connections between the actions and how the technical artefact is supposed to respond to them. Finally, designing a user interface relies on the ways in which the artefact or system is used in an action. Task analyses offer one perspective on interaction. They explain how different action goals and sub-actions and their functional goals are reached by using technical artefacts. Event flows and event trees describe how users can control technical artefacts from one state to another. Controlling the artefact or system is dependent on the control tools and instruments that are used to reach certain goals. These tools as a whole form the user interface, which comprises the required control equipment and different interaction modes.

In event trees, the role of the user interface is decisive when users make decisions about the future course of actions. Technical artefacts are unable to set goals for an action or choose between different possible goals. Only human users or designers can decide what the relevant goals are, and user interfaces can provide them with different opportunities to make choices. Providing the scope to set goals is the main function of the user interface, as this enables users to control the technology.

One can sometimes see solutions in which users are required to interact with a technical process in a way that leaves them no real choices. For example, their task may be to make the artefact move from one state to another to generate the next step in an event path. An example of such a bad design is the extensive use of unnecessary message boxes (Cooper et al. 2007). While in some cases confirmations of user actions provided by message boxes make sense, they are simply meaningless and irritating if they are used only to confirm the physical processes of the device. In a good design, a user's actions should only be needed to select between different critical goal states for an artefact.

Rational selections presuppose valid information about the given situation. In order to succeed in their tasks, users need meaningful feedback about the internal and external states of the technical artefact. Feedback instruments provide users with the information they need to make decisions regarding the control of a technical artefact.

Concrete human actions in using a technical artefact can be carried out by using input/output (I/O) components (Cooper et al. 2007; Griggs 1995; Henry 1998). These devices are used to control the behaviour of an artefact. They are composed of many kinds of objects, devices, and tools that allow users to regulate the behaviour of technologies and activate relevant I/O operations (Greenstein and Arnaut 1988; McKay 1999; Salvendy 2006). They must be connected to instruments that indicate the currently prevailing states and contextual factors of the technical artefact so the user can control its behaviour in a sense-making and logical manner.

These types of control can be divided into two main classes. First, there are instruments that are used to input information into the system. Keyboards, pointing devices, and touchscreens are typical examples of such instruments in modern computer-based technologies, but one can also base input information, for example, on movement recognition, thermal energy, biometric characteristics, haptic signals, pressure and voice, and even feed the information remotely with the help of another device, such as a smartphone (Cooper et al. 2007; McKay 1999). Traditional machine controls—such as steering wheels, gears, levers, push buttons, and foot pedals—are also equally important devices for controlling certain artefacts. Indeed, there are many ways to enter information into a device in order to control it and reach the expected goal state.

The second class of controls comprises instruments used for output purposes—such as displays, meters, pop ups, warning signals, heads up displays (HUDs), and vibrations—to deliver information about the state of the device or system and its related context, such as weather conditions or noise. Such controls provide information required to assess important control dimensions.

There are typical input and output components for GUI (Cooper et al. 2007; McKay 1999), some of which are designed to input information into the artefact (such as text boxes or forms), and some that have been developed to directly control the behaviour of the programs. Typical examples of the latter include dialogue boxes, gizmos (or controls), scrollbars, radio buttons, menus, and toolbars (Cooper et al. 2007; McKay 1999). The main advantages of such standardized interface components are the positive transfer effects created by the consistency of user interfaces and the ease of learning how to operate new interfaces (Helfenstein and Saariluoma 2006; Singley and Anderson 1987). Another benefit can be found in programming interfaces: standardized parts allow the wide reuse of codes and make visual programming easier.

Thus user interfaces can be based on different technological solutions, which can be used to classify them into different types. Some user interfaces are mechanical (such as a steering wheel, a gear, or a keyboard), while others are based on commands as in computer languages and still others are graphical such as modern web pages. Some interfaces are based on such ubiquitous interaction solutions as the direct registration of a human body and its neural states (Weiser 1993).

These kinds of interface categories are useful but not absolute, because graphical interfaces can have mechanical features, and mechanical interfaces may have electromechanical and graphical features. For instance, touchscreen interfaces have mechanical aspects (the user presses a surface), and power steering systems are hydraulic and electronically realized, whereas steering wheels are mostly mechanical devices. Hence, from the users' point of view, the discussion of the technical classification of user interfaces is of limited importance when constructing theoretical concepts for user interface design.

Dialogues and Interaction Semiotics

The HTI process entails many forms of interaction. Only completely automated systems can be operated by turning the system on using a single command, after which the system takes care of everything else. Yet, even these systems ultimately depend on the initial human action (Minsky 1967). User interfaces that can be operated via a single command are the ideal form of interaction, but unfortunately this is increasingly rare in modern information technology.

Technical artefacts and goods have become increasingly complex. People are required to operate using many kinds of input modes and hierarchies when performing a task, and users have to control event flows in different states of operation. These tasks are usually carried out in the form of a dialogue or a game between the user and the artefact. Designers must create sense-making dialogues to enable the artefact's effective use (Cooper et al. 2007; Moran 1981).

In constructing dialogues, it is essential to provide users with as much relevant knowledge as possible to help them choose between different actions. This presupposes goal-relevant decisions during the dialogue. Designers use task analysis to understand what people intend to achieve during different sub-tasks in order to define how the given technology should respond to users' actions and what kind of feedback should be developed in the user interface for different inputs.

User interfaces are sign systems. They may provide information about many different types of issues, and users are expected to comprehend the delivered messages. The sub-set of semiotics, the study of signs, which relates to user interfaces and other HTI and communication issues, is called semiotic engineering (de Souza 1993, 2005; Rousi 2012). In the following, we examine the overall principles of semiotics and semiotic systems of user interfaces.

The roots of semiotics (Eco 1976) can be found in the classic works of Peirce (1931–1958) and de Saussure (1916). The key problem of semiotics is using a sign to stand for something else, such as opening a web page (i.e., a path in an event tree). Signs have the capacity to evoke mental representations of references in the human mind. They can be words or linguistic signs; they can be formal signs such as mathematical symbols, or commands in computer languages. They can also be natural phenomena such as smoke, which signifies fire, or fever, which signifies illnesses (Eco 1976). They also provide us with a valuable conceptual tool for investigating and designing interfaces.

Semiosis is the process by which people give meanings to signs. The context of giving meanings can be called semiosphere or a language game (Lotman 2005; Wittgenstein 1953). These concepts refer to how the meanings of signs are defined in a particular discourse—such as a user interface system—which provides valuable tools for interface designers. All user interfaces must have their own system of signs, though within a single interface paradigm the same symbol or sign may have different meanings.

An example of applying semiotic concepts to analyse the communication between technical artefacts and users is Peirce's triadic analysis of sign types (Peirce 1931–1958). Peirce divided signs into three types: icons, indexes, and symbols; this differentiation is useful in interface semiotics. Icons are pictures that resemble a reference, whereas indexes include a direct connection to the reference. Thus, many natural phenomena can serve as alarms, such as smoke to warn of fire or mercury to indicate high or low temperature. Similarly, a picture can represent a person as an icon. Finally, symbols stand for their references. Thus words can have a symbolic value in interfaces.

It is important to create logical and coherent semiospheres in interface design, which the user can effectively use to control technical artefacts. Overall principles of semiotics are needed for interaction languages (de Souza 2005), as well as knowledge of their operation in the mind. Ogden and Richards (1923) argue that in human thoughts, signs have their own references, which refer to referents (external objects, events, or ideas). For example, the word 'dog' refers to the idea of a dog in the human mind, and this idea refers to an actual dog. Signs thus acquire their meanings via an act of the human mind. Accordingly, technical signs can be constructed in the context of the human mind and its laws, which requires empirical psychological justification of the meaning of signs.

Research on mental processes that give meaning to technical signs (e.g., signs in technical artefacts) can be called technological psychosemiotics (Saariluoma and Rousi 2015). This paradigm investigates how people give meaning to technical signs such as icons, controls, or meters by using psychological and socio-cultural research methods, concepts, and theories. It analyses signs in technical artefacts to provide designers with objective knowledge of the relevant semiotic systems.

Design Patterns

Many existing components are used in current technology development. Game programming, for example, relies on previous programs and their components. In most game development the story and the design of the figures are more demanding than the actual programming work, which employs several standard user interface components, technology standards, and styles (Cooper et al. 2007; McKay 1999). Standardized design solution patterns are also commonly used in the physical interaction designs that are typical of mechanical technologies (Alexander 1977). Standardized solution patterns also make it easier for users to learn and accept new systems.

Mechanical interfaces have standard input and output components (Dieter and Schmidt 2009). Some of them are used to control the spatial position of moving objects (such as a steering wheel or a rudder), while others (such as an accelerator or brake pedal) are used to control the speed of a process. It is also essential to have information about the internal and external states of a given artefact in order to use the technology. This information cannot always be detected by the human senses. The temperature of combustion processes, for example, is not definable by human senses. Similarly, it is impossible to be aware of the thousands of indicators that ensure the proper paper-making processes in a modern paper machine. Only with the help of adequate user interface components is it possible for users to supervise the processes of machines.

Many standard control components have been developed in mechanical engineering in recent decades. Such items as hand wheels, knobs, levers, joysticks, rocker switches, pedals, push buttons, and sensor keys have been used for a long time in technical interaction solutions as input controls for machines. Their design parameters are well known and their usage has frequently been tested.

Likewise, there are also numerous more or less standardized output instruments, such as speedometers, thermometers, and artificial horizons. Modern paper machines, for example, can have over 4000 sensors registering different aspects of this complex process. Successfully integrating these sensors (and the feedback they give controllers) is crucial for keeping these complex systems effectively operable.

Standardized interaction components can also be found in the area of HCI (Cooper et al. 2007; Griggs 1995; Goodwin 2011). Event handlers can steer computational processes on the basis of information that users have directly or indirectly input into the computer. Such standard elements as text and dialogue boxes, radio and option buttons, toolbars, icons, pop ups, drop-down menus, or scrollbars are everyday components used to build standard GUIs (Cooper et al. 2007; Griggs 1995; McKay 1999). Feedback is mostly given on a screen, but it can also be auditory or printed. The main function of feedback is to provide users with information to help control the process or support the task.

In the design, it is necessary to anticipate the usage situation and the related requirements in order to develop the best possible means for the interaction. Depending on the situation, the solution can involve anything from hand and foot controls and visual and auditory displays to designing suitable illumination conditions and organizing the space for effective use of the given input tools (Sanders and McCormick 1993). To approach the questions of user interface design systematically, the ICT design community has developed a set of recommendations, guidelines, technology standards, and de facto standards for user interfaces (Cooper et al. 2007; McKay 1999). These mainly discuss interface components that have been widely applied in user interface design.

Usability standards and style guidelines also support reusing design ideas. Standards are intended as norms to help designers unify the technical culture; they are often worldwide requirements that designers follow globally. The content of standardization of HTI issues has improved along with the recent ISO-9241 standards, which include more specific information about ergonomics, safety and interaction issues than earlier versions (ISO 1998a, b). There are also laws and directives that regulate the technical aspects of interaction. These are mostly technical details about health risks, such as the influence of magnetic fields on pacemakers.

Though the above-mentioned guidelines and directives provide support information for constructing technical HTI processes, designers must decide between alternative ways of creating interaction solutions (or develop their own variations). This leaves room for many style standards for web pages, for example. Style sheets have standardized many properties of web pages, for example typical interaction components such as fonts, boxes, lists, and positioning have been given their own requirement recommendations (Collison et al. 2009; w3schools 2013). There are also a huge number of templates and other ready-to-modify solution models for solving concrete design problems.

Standardized solution models simplify the design of technical solutions and interfaces. Programmers and mechanical designers can use the same solutions again, which saves considerable time and facilitates a more unified and easy-to-learn design culture. However, in most cases standard solutions and best practices offer only a framework within which one can work, rather than a final solution. In the design, it is essential to understand why given standard solutions and standards make sense, as well as why they are not optimal in some situations.

The Conceptual Structure of 'How to Control'

In summary, the first level of HTI design is technical, and is based on traditional technical engineering concepts (Eder and Hosnedl 2008; Pahl et al. 2007; Ulrich and Eppinger 2011). Here, the main problem is to create a technical artefact that can carry out a given task defined by a *user need*. For example, a drill is designed to make a hole in a certain material. In order to get technical artefacts to do what they should to support users in their actions, it is essential to provide users with the means to control them. This aspect of functionality design is called *technical user interface* construction.

Based on knowledge of people's intentions and goals, it is possible to define the necessary *functionalities* of the technology (Pahl et al. 2007; Ulrich and Eppinger 2011). Designing functionalities plays a central role in all engineering design, as every part of a system has its own function in the whole (Boyle et al. 2013; Dym and Brown 2012; Gero 1990; Goel et al. 2009; Hirtz et al. 2002). In HTI design, functionality defines how a technology can improve a particular human action (i.e., what people can do with the given technical artefact). Doctors, for example, need knowledge about their patients in order to make diagnoses or safeguard their living environments. By using correctly designed information systems, medical staff can acquire a more general view of their patients' medical histories than can be provided in a patient interview (Barach and Small 2000; Blumenthal 2010).

The 'how to control' problem has a special conceptual structure. So far, we have discussed its main components, but a more holistic picture of the process provides a useful context. The overall structure of human interaction with technical artefacts is similar from one device to another, since the overall structure of all artefacts that are controlled by people is basically the same (Minsky 1967; Turing 1936–1937). The *how-to-control discourse* opens up four important sub-discourses. First, one has to understand users' tasks when a new technology is being used. Second, it is necessary to comprehend the functions of the technology in performing the task in question. Third, it is essential to understand the context of using the technology. Fourth, one has to construct the behaviour of the artefact and how users can control it. Each of these questions must be discussed before proceeding further in HTI design.

When the goals of a technical artefact (and the processes for reaching those goals) are determined, the next step is to define what users must do to reach the goals. Technical artefacts have a finite set of input channels, which in turn have a finite set of possible states. Analogical control devices such as steering wheels can be 'digitalized' and interpreted as finite state machines because their principal infinity does not have practical meaning as they can be digitalized similarly to digital music.

A technical artefact has also a finite *set of output channels*, which in turn have a finite set of possible states for each channel (Minsky 1967; Turing 1936–1937). Using technology is based on the users' operations, which control the behaviour of the artefact and the systems feedback. To summarize, the overall conceptual structure of the process of using any technical artefact is as follows: After defining the overall goal of the user, one must define the goals for (and the functionalities of) using the technical artefact or system to carry out the user's specific tasks and sub-tasks. This can be expressed by defining the expected goal states of the given technology and the respective technical processes from its initial or prevailing state to the final goal state. It is also necessary to define the input channels that enable people to control the behaviour of the artefact so it can reach its goal state. Finally, it is essential to define the technical and natural feedback systems that allow users to follow the artefact's behaviour.

All artefacts have this type of conceptual structure. It defines the fundamental questions associated with the problems of how-to-control technologies, which must be solved in any HTI design project.

The main concepts and design problems of how-to-control are:

- The main aim of the action (the user's goal) must be defined.
- The technology's effect on the environment (natural, mental, information, or socio-cultural) must be determined.
- The user's tasks and sub-tasks must be defined.
- The technical artefact or system is transformed from its current initial state to its expected goal state to achieve the effect.
- Users employ tasks when using a technology and each task has multiple sub-tasks.
- All tasks and sub-tasks have a purpose in human action when using the technical artefact or system.
- Each technical artefact or system has input channels with different operational states.
- Each element of user input affects the behaviour of the technical artefact or system.
- Each technical artefact or system has a number of output channels with multiple output states that give the user feedback information about the prevailing state of the artefact.
- Natural observation (following how the artefact behaves) serves as an output channel.
- The combination of input and output channels constitutes the human–artefact interface.
- A user interface must be implemented to direct the actions of the artefact.

Before defining the contents and functions of information channels, it is essential to assemble a complete representation of the control of technology, which requires considering the kinds of operations and technical goals the artefact is supposed to achieve in support of overall human goals, and the contexts in which it will be used. I/O component standards and style guides are valuable and necessary in technical design. However, they are not sufficient for understanding and designing HTI processes. Before using them, it is essential to consider which standards and guidelines are justifiable (and why) by proceeding from technical discourse to psychological and sociological discourse and examining concepts of human research in the field of technology design. Technological standards and design solutions must be justified by their positive effects on human physical and biological environments, mentality, and society.

References

Alexander, C. (1977). *A pattern language: Towns, buildings, construction.* Oxford: Oxford University Press.

Annett, J. (2000). Theoretical and pragmatic influences on task analysis methods. In J. Schraagen, S. Chipman, & V. Shalin (Eds), Cognitive Task Analysis (pp. 25–40). Malwah, NJ: Erlbaum.

Annett, J. (2004). Hierarchical task analysis. In D. Diaper & N. Stanton (Eds.), *Handbook of cognitive task design* (pp. 63–82). Hillsdale, NJ: Erlbaum.

Annett, J., & Duncan, K. D. (1967). Task analysis and training design. *Report resumes.* Hull: Hull University.

Barach, P., & Small, S. D. (2000). Reporting and preventing medical mishaps: Lessons from non-medical near miss reporting systems. *British Medical Journal, 320,* 759–763.

Blumenthal, D. (2010). Launching HITECH. *New England Journal of Medicine, 362,* 382–385.

Bouma, H., Fozard, J. L., & van Bronswijk, J. E. M. H. (2009). Gerontechnology as a field of endeavour. *Gerontechnology, 8,* 68–75.

Boyle, E., Van Rosmalen, P., & Manea, M. (2013). *Cognitive task analysis.* Retrieved April 23, 2015, from http://dspace.learningnetworks.org/bitstream/1820/4848/1/CHERMUG-Deliverable%2014-CognitiveTaskAnalysis-WP2.pdf

Brentano, F. (1874/1955). *Psychologie vom empirischen Standpunkt.* Hamburg: Felix Meiner.

Card, S., Moran, T., & Newell, A. (1983). *The psychology of human-computer interaction.* Hillsdale, NJ: Erlbaum.

Chandrasekaran, B. (1990). Design problem-solving—A task-analysis. *Ai Magazine, 11,* 59–71.

Charness, N. (2009). Ergonomics and aging: The role of interactions. In I. Graafmans, V. Taipale, & N. Charness (Eds.), *Gerontechnology: Sustainable investment in future* (pp. 62–73). Amsterdam: IOS Press.

Collison, S., Budd, A., & Moll, C. (2009). *CSS mastery: Advanced web standards solution*. Berkeley, CA: Friends of ED.

Cooper, A., Reimann, R., & Cronin, D. (2007). *About Face 3: The essentials of interaction design*. Indianapolis, IN: Wiley.

Crandall, B., Klein, G., & Hoffman, R. R. (2006). *Working minds: A practitioner's guide to cognitive task analysis*. Cambridge, MA: MIT Press.

Czaja, S. J., & Nair, S. N. (2006). Human factors engineering and systems design. In G. Salvendy (Ed.), *Handbook of human factors and ergonomics* (pp. 32–49). Hoboken, NJ: Wiley.

Datye, S. (2012). Life-based design for technical solutions in social and voluntary work. In *Jyväskylä studies in computing* (Vol. 164). Jyväskylä: University of Jyväskkylä Press.

de Saussure, F. (1916/2011). *Course in General Linguistics*. New York, NY: Columbia University Press.

de Souza, C. S. (1993). *The semiotic engineering of user interface languages*. International Journal of Man-Machine Studies, 39, 753–773.

de Souza, C. S. (2005). *The semiotic engineering of human-computer interaction*. Cambridge, MA: MIT Press.

Dennett, D. (1991). *Consciousness explained*. Boston, MA: Little Brown.

Diaper, D. (2004). Understanding task analysis for human-computer interaction. In D. Diaper & N. Stanton (Eds.), *The handbook of task analysis for human-computer interaction* (pp. 5–47). Mahwah, NJ: Erlbaum.

Dieter, G. E., & Schmidt, L. C. (2009). *Engineering design*. Boston, MA: McGraw-Hill.

Dym, C. L., & Brown, D. C. (2012). *Engineering design: Representation and reasoning*. New York: Cambridge University Press.

Eco, U. (1976). *A theory of semiotics*. Bloomington, IN: Indiana University Press.

Eder, W., & Hosnedl, S. (2008). *Design engineering. A manual for enhanced creativity*. Boca Raton, FL: CRC Press.

Gaver, W. W. (1991). Technology affordances. In *Proceedings of the SIGCHI Conference on Human Factors in Computing Systems* (pp. 79–84).

Gero, J. S. (1990). Design prototypes: A knowledge representation schema for design. *AI Magazine, 11*, 26–36.

Gero, J. S., & Kannengiesser, U. (2004). The situated function–behaviour–structure framework. *Design Studies, 25*, 373–391.

Gero, J. S., & Kannengiesser, U. (2014). The function-behaviour-structure ontology of design. In A. Chakrabarti & L. Blessing (Eds.), *An anthology of theories and models of design* (pp. 263–283). London: Springer.

Gibson, J. J. (1979). *The ecological approach to visual perception.* Boston, MA: Houghton Mifflin.

Go, K., & Carroll, J. M. (2004). The blind men and the elephant: Views of scenario-based system design. *Interactions, 11,* 44–53.

Goel, A. K., Rugaber, S., & Vattam, S. (2009). Structure, behavior, and function of complex systems: The structure, behavior, and function modeling language. *Artificial Intelligence for Engineering Design, Analysis and Manufacturing, 23,* 23–35.

Goodwin, K. (2011). *Designing for the digital age: How to create human-centered products and services.* Indianapolis, IN: Wiley.

Greenstein, J., & Arnaut, L. (1988). Input devices. In M. Helander (Ed.), *Handbook of human-computer interaction* (pp. 495–519). Amsterdam: North-Holland.

Griggs, L. (1995). *The windows interface guidelines for software design.* Redmond, WA: Microsoft Press.

Helfenstein, S., & Saariluoma, P. (2006). Mental contents in transfer. *Psychological Research, 70,* 293–303.

Henry, P. (1998). *User-centred information design for improved software usability.* Boston, MA: Artech.

Hirtz, J., Stone, R. B., McAdams, D. A., Szykman, S., & Wood, K. L. (2002). A functional basis for engineering design: Reconciling and evolving previous efforts. *Research in Engineering Design, 13,* 65–82.

Hollnagel, E. (2006). Task analysis: Why, what, and how. In G. Salvendy (Ed.), *Handbook of human factors and ergonomics* (pp. 371–383). Hoboken, NJ: Wiley.

Husserl, E. (1901–1902). *Logische unterschungen* (Vols. I–II). Halle: Niemeyer.

Hysong, S. J., Sawhney, M. K., Wilson, L., Sittig, D. F., Esquivel, A., & Singh, S., et al. (2011). Understanding the management of electronic test result notifications in the outpatient setting. *BMC medical Informatics and Decision Making, 11.* Retrieved February 28, 2015, from http://www.biomedcentral.com/1472-6947/11/22

International Organization for Standardization. (1998a). *ISO 9241-11: Ergonomic Requirements for Office Work with Visual Display Terminals (VDTs): Part 11: Guidance on Usability.*

International Organization for Standardization. (1998b). *ISO-14915: Ergonomic Requirements for Office Work with Visual Display Terminals (VDTs): Part 11: Guidance on Usability.*

Karwowski, W. (2006). The discipline of ergonomics and human factors. In G. Salvendy (Ed.), *Handbook of human factors and ergonomics* (pp. 3–31). Hoboken, NJ: Wiley.

Kieras, D. E., & Meyer, D. E. (1997). An overview of the EPIC architecture for cognition and performance with application to human-computer interaction. Human-Computer Interaction, 12, 391–438.

Kjeldskov, J., & Paay, J. (2012). A longitudinal review of Mobile HCI research methods. In *Proceedings of the 14th International Conference on Human-Computer Interaction with Mobile Devices and Services* (pp. 69–78). ACM. Retrieved February 28, 2015, from http://people.cs.aau.dk/~jesper/pdf/conferences/Kjeldskov-C65.pdf

Leikas, J. (2009). *Life-based design—A holistic approach to designing human-technology interaction.* Helsinki: Edita Prima Oy.

Lin, J., Newman, M. W., Hong, J. I., & Landay, J. A. (2000). DENIM: Finding a tighter fit between tools and practice for web site design. In *Proceedings of the SIGCHI Conference on Human Factors in Computing Systems* (pp. 510–517).

Lotman, Y. (2005). Semiosphere. *Sign-Systems Studies, 1,* 205–229.

Markopoulos, P., & Bekker, M. (2003). Interaction design and children. *Interacting with Computers, 15,* 141–149.

McKay, E. (1999). Exploring the effect of graphical metaphors on the performance of learning computer programming concepts in adult learners: A pilot study. *Educational Psychology, 19,* 471–487.

Memon, A. M., Soffa, M. L., & Pollack, M. E. (2001). Coverage criteria for GUI testing. *ACM SIGSOFT Software Engineering Notes, 26*(5), 256–267.

Minsky, M. L. (1967). *Computation: Finite and infinite machines.* Englewood Cliffs, NJ: Prentice-Hall.

Moran, T. P. (1981). Guest editor's introduction: An applied psychology of the user. *ACM Computing Surveys, 13,* 1–11.

Morgan, D. L. (1996). *Focus groups as qualitative research.* Thousand Oaks, CA: Sage.

Myers, B., Hudson, S. E., & Pausch, R. (2000). Past, present, and future of user interface software tools. *ACM Transactions on Computer-Human Interaction (TOCHI), 7,* 3–28.

Newell, A., & Simon, H. A. (1972). *Human problem solving*. Engelwood Cliffs, NJ: Prentice-Hall.

Newman, M. W., & Landay, J. A. (2000). Sitemaps, storyboards, and specifications: A sketch of web site design practice. In *Proceedings of the 3rd Conference on Designing Interactive Systems: Processes, Practices, Methods, and Techniques* (pp. 263–274).

Nickerson, R., & Landauer, T. (1997). Human-computer interaction: Background and issues. In M. Helander, T. Landauer, & P. Prabhu (Eds.), *Handbook of human-computer interaction* (pp. 3–31). Amsterdam: Elsevier.

Nielsen, J. (1993). *Usability engineering*. San Diego, CA: Academic Press.

Ogden, C., & Richards, I. (1923). *The meaning of meaning*. London: Routledge and Kegan Paul.

Pahl, G., Beitz, W., Feldhusen, J., & Grote, K. H. (2007). *Engineering design: A systematic approach*. Berlin: Springer.

Peirce, C. S. (1931–1958). In C. Hartshorne, P. Weiss, & A. Burks (Eds.), *Collected papers of Charles Sanders Peirce* (Vols. 1–8). Cambridge, MA: Harvard University Press.

Rasmussen, J., Mark Pejtersen, A., & Goodstein, L. P. (1994). *Cognitive systems engineering*. New York: Wiley.

Rauterberg, M. (1996). How to measure the ergonomic quality of user interfaces in a task independent way. In A. Mital, H. Krueger, S. Kumar, M. Menozzi, & J. E. Fernandez (Eds.), *Advances in occupational ergonomics and safety I* (pp. 154–157). Cincinnati, OH: International Society for Occupational Ergonomics and Safety.

Rosson, B., & Carroll, J. (2002). *Usability engineering: Scenario-based development of human-computer interaction*. San Francisco, CA: Morgan Kaufmann.

Rousi, R. (2012). From cute to semiotics.

Saariluoma, P., & Rousi, R. (2015). Symbolic interactions: Towards a cognitive scientific theory of meaning in human technology. *Journal of Advances in Humanities, 3*, 310–323.

Salvendy (2006), is editor of G. Salvendy (Ed.), In Handbook of Human Factors and Ergonomics. Hoboken, NJ: John Wiley & Sons.

Sanders, M. S., & McCormick, E. J. (1993). *Human factors in engineering and design* (7th ed.). New York: McGraw-Hill.

Searle, J. (1992). *The rediscovery of mind*. Cambridge, MA: MIT Press.

Shneiderman, B. (1983). Direct manipulation: A step beyond programming languages. *IEEE Computer, 16*, 57–69.

Shneiderman, B., & Maes, P. (1997). Direct manipulation vs. interface agents. *Interactions, 4*, 42–61.

Singley, M. K., & Anderson, J. R. (1987). A keystroke analysis of learning and transfer in text editing. *Human-Computer Interaction, 3*, 223–274.

Smith, B. (2003). The logic of biological classification and the foundations of biomedical ontology. In *Invited Papers from the 10th International Conference in Logic Methodology and Philosophy of Science* (pp. 19–25). Oviedo, Spain.

Stanton, N. A. (2006). Hierarchical task analysis: Developments, applications, and extensions. *Applied Ergonomics, 37*, 55–79.

Taylor, F. (1911). *Shop management.* New York: McGraw-Hill.

Taylor, C. (1964). *The explanation of behaviour.* London: Routledge and Kegan Paul.

Turing, A. M. (1936–1937). On computable numbers, with an application to the entscheidungsproblem. *Proceedings of the London Mathematical Society, 42*, 230–265.

Ulich, E., Rauterberg, M., Moll, T., Greutmann, T., & Strohm, O. (1991). Task orientation and user-oriented dialog design. *International Journal of Human-Computer Interaction, 3*, 117–144.

Ulrich, K. T., & Eppinger, S. D. (2011). *Product design and development.* New York: McGraw-Hill.

Vicente, K. J. (1999). *Cognitive work analysis: Toward safe, productive, and healthy computer-based work.* Mahwah, NJ: Erlbaum.

von Wright, G. H. (1971). *Explanation and understanding.* London: Routledge and Kegan Paul.

w3Schools. (1999–2016). Retrieved February 12, 2011, from http://www.w3schools.com/css/

Weiser, M. (1993). Some computer science issues in ubiquitous computing. *Communications of the ACM, 36*, 75–84.

Wittgenstein, L. (1953). *Philosophical investigations.* Oxford: Basil Blackwell.

Woods, D. D., & Roth, E. M. (1988). Cognitive engineering: Human problem solving with tools. *Human Factors: The Journal of the Human Factors and Ergonomics Society, 30*, 415–430.

Yuan, X., Cohen, M. B., & Memon, A. M. (2011). GUI interaction testing: Incorporating event context. *IEEE Transactions on Software Engineering, 37*, 559–574.

4

The Psychology of Fluent Use

In a perfect world, it would always be possible to operate technology effortlessly and to reach the desired goal. However, in the real world many factors may make technologies difficult to use or even hinder people from using technical artefacts. Most of these factors pertain to usability (i.e., technology's ability to fit users' capabilities) and thus concern technological solutions from the point of view of human beings as users of technology. Therefore, designing technical artefacts that are easy to use requires understanding the psychological and mental preconditions for using technology.

People have a variety of different functionalities in their smartphones and laptops that they never use although they would need them in their life (Kämäräinen and Saariluoma 2007; Shneiderman 2011). There might be attractive applications and services that users could benefit from, but they do not know what they are or how to use them. They cannot access the devices or systems in the intended way, and do not know how to navigate or input information. They may also have special requirements for accessing technology due to physical restrictions, or quite simply a lack of training (Blackmon et al. 2005; Dillon et al. 1990). Failing to use technology in an adequate manner can even lead to risky interaction situations, examples of which can be found in traffic and industrial accidents (Lamble

© The Editor(s) (if applicable) and The Author(s) 2016
P. Saariluoma et al., *Designing for Life*,
DOI 10.1057/978-1-137-53047-9_4

et al. 1999; Laursen et al. 2008). In principle, people should be able to use all the facilities and functionalities that technologies offer, but in practice this is not always the case. This is one of the main reasons that products fail to find markets (Norman 1999; Shneiderman 2011). Designing relevant and attractive functionalities for technologies is important, but it is even more necessary to ensure that people can really use them.

User interface solutions are meant to provide people with a realistic *possibility* of reaching their goals (Card et al. 1983; Nielsen 1993; Rosson and Carroll 2002; Shneiderman, and Plaisant 2005; Wickens and Holands 2000), regardless of users' capacity limitations (Baddeley 2007, 2012; Broadbent 1958; Covan 2000; Miller 1956), comprehension problems (Kitajima and Polson 1995), or lack of skills (Green and Petre 1996; Navarro-Prieto and Canas 2001; Visser and Hoc 1990). Therefore, we focus the design discourse now on the issues how to make technologies easier for people to use (Card et al. 1983; Olson and Olson 2003; Rosson and Carroll 2002).

The ability to use technology depends on mental processes (Karwowski 2006). The issues of 'to be able to' and 'ease of use' are typical problems of usability, accessibility, ergonomics, and human factors (Bridger 2009; Karwowski 2006; Kivimäki and Lindström 2006; Rosson and Carroll 2002; Nielsen 1993). In addition to these research domains, which are based mainly on natural scientific laws, understanding the characteristics of the human mind is essential when designing non-functional human requirements for technical artefacts.

The next question is thus how to apply the knowledge of psychology and other related human sciences in designing HTI solutions. This knowledge is often analytical, in the sense that it is composed of detailed pieces of information from different domains of human sciences. This information, along with many different sub-problems, forms a holistic approach to the human dimension of technology.

Critical Factors of 'Being Able to Use'

The core questions of 'being able to use' in HTI can be found mostly in the area of psychology (Card et al. 1983), in particular user psychology (Moran 1981; Saariluoma 2005; Saariluoma and Oulasvirta 2010), supported by other human sciences, such as sociology and brain physiology (Cockton 2004; Parasuraman and Rizzo 2006; Pinch and Bijker 1984; Sassen 2002). User psychology, a branch of psychological thinking, examines the problems of using technology in light of psychological concepts, knowledge, theories, and methods. It is thus comprised of both scientific knowledge and the practical problems of using technology.

A common way of applying psychological knowledge is to articulate typical questions related to HTI and to associate the respective psychological research with each critical problem. Defining general user interaction problems can do this. Figure 4.1 (cf. Card et al. 1983) illustrates a human user interacting with a technical artefact.

Ultimately, all artefacts have identical basic interaction properties and a conceptual structure to the extent that, after receiving input information from the user, the artefact interacts with the environment and gives feedback to the user. This logic follows the concept of a Turing (1936–1937) machine (Minsky 1967). By analysing the crucial psychological

Fig. 4.1   Using a technical artefact

sub-tasks in this process, it is possible to define the basic design questions of any technology.

In this way, it would be possible to apply psychological knowledge on a general level in considering the issue of 'being able to use', and to ask what kinds of psychological factors may best support the interaction and what kind of psychological limitations of the human mind may make it difficult (or even impossible) to interact with technologies.

A common-sense analysis of the interaction situation generates a number of psychologically motivated sub-tasks that the user must be able to carry out. First, the user must be aware of the present stage of the task and its future course. He or she must also be able to detect the information given by the artefact about its internal and external states. These *perceptual discrimination problems*, which are discussed in the psychology of perception and attention (Covan 1999, 2000; Pashler 1998; Styles 1997), must be solved to give the user the relevant information that is needed to operate the artefact (Proctor and Proctor 2006).

The second fundamental task for the user is how to control the behaviour of the artefact by communicating information about the user's goals to the artefact. Today, controlling technical artefacts is mainly a physical task, which requires motor actions. These actions can be carried out, for example, by hand, feet, or body movements, or by looking, speaking, or using the vibrations of the larynx. The user's motor actions—and the optimal conditions for carrying these actions out—comprise a central psychological and usability problem in design, as psychomotor behaviour and perceptual discrimination are intimately connected in a perceptual-motor loop, which forms an essential part of human cognition (Jeannerod 2007). This loop is important in ubiquitous computing, as well as in pervasive or ambient computing (Weiser 1993).

Using artefacts requires the user to perform an organized series of tasks and store large chunks of relevant information in his or her immediate memory to be able to carry out the tasks. In addition to the immediate perceptual and motor cycles, the user has to be able to keep active representations of more complex tasks in his or her mind. The *working memory and remembering* thus form the next problem of usability design. The relevant psychological concepts are short- and long-term working memories, as these two systems keep task-relevant information active in the human mind (Baddeley 2007, 2012; Chase and Ericsson 1981; Covan 1999, 2000; Ericsson 2006; Gopher and Donchin 1986).

Some alternatives to the Ericsson and Kintsch (1995) model have been suggested in literature. Alternative views have been presented by Baddeley (2007). His model is called episodic long-term memory. Another alternative has been suggested by Gobet (2000). Nevertheless, all models imply that large active mental representations must be partly deposited in more permanent storage than the easily inferable working memory. This storage component is better known in terms of what Ericsson and Kintsch (1995) describe as long-term working memory (LTWM). From a design point of view, the important thing is that training makes it possible for people to activate sizable mental representations and store them in memory for longer periods of time and in larger amounts than predicted from the traditional working memory models. Thus, LTWM explains why some actively used materials are beyond standard interference. One can use this storage to analyse and explain phenomena such as interruptions, remembering computer program code, and learning to use computers (Ericsson and Kintsch 1995; Oulasvirta and Saariluoma 2004, 2006).

Using technical artefacts does not only involve perceiving and responding to the operation of the artefact. Users are also required to learn which responses are relevant in which situations. This can be called the problem of perceptual-motor learning (or in a wider sense, the problem of learning to use technologies). The use of most technical artefacts requires some learning, and sometimes a learning process can be very complicated and demanding, for example, programmers must study for years to reach the required skill level (Anderson et al. 1984; Mayer 1997; Robins et al. 2003).

Acquiring expertise in using advanced technologies often takes a long time, and even experts must keep in mind the issues they have learned (Mayer 1997). This brings about the next interaction task, which can be called *long-term remembering*; the psychological ability that is responsible for it can be called long-term memory. It is a vital concept in learning. Everything that has been learned is kept in the long-term memory (Baddeley 2007, 2012; Brady et al. 2009).

Additional tasks must be studied before one can draw a road map of the psychological sub-tasks associated with using a technical artefact.

The next task is to *organize communication* so that it is comprehensible. The user has to understand the communication between herself and the technical artefact. At best, this is intuitive or self-evident, and the communication is organized in a way that it is easily comprehensible. Thus, the psychological problem of *comprehension* is important in HTI (Brooks 1980; Kintsch and van Dijk 1978; Pennington 1987; Salmeron et al. 2005).

Comprehension, however, is not sufficient to help the user complete the task. He or she must also be able to decide between different alternative actions. Therefore, *decision making* is another important task for the user (Gigerenzer and Gaissmaier 2011; Kahnemann 2011). The more complex the technology, the more essential this task is. It can even be critical in the usage of complex technologies, such as power stations or financial investment software, where the user has to be able to make up his or her mind concerning different courses of actions (Lehto and Nah 2006), and where the quality of decisions may entail, for example, security and economic risks.

The psychology of decision making is essential in investigating the decision aspects of HTI (Gigerenzer and Gaissmaier 2011; Kahnemann 2011; Simon 1955). The tasks that people are expected to execute using today's sophisticated technologies are increasingly complex and presuppose complex problem-solving skills. *Situation awareness* entails problem solving that relies on consciousness of the present situation (Endsley 1995, 2006; Endsley and Jones 2011). In a risk society—a consequence of the increasing complexity of technologies (Beck 1992, 2008)—critical situations, such as nuclear accidents, require human capability to take technical decisions rapidly and accurately in order to minimize harm to people. Investigating the nature of complex problem solving (and thereby the role of psychological interaction design) has become more important in the field of HTI (Anderson 1983, 1993; Funke 2012).

Interaction with technologies is not only a cognitive issue; it has emotional, motivational, and personal aspects as well (Hassenzahl 2011; Norman 2004). Factors such as *emotional processes*, *motivation*, and *personality* are present in the interaction and must be included in user psychological research and HTI design processes. These issues will be discussed in the following chapter. However, it is not only tasks that differ from each other. Users of technology are also different. They may have different societal, occupational, educational, and generational backgrounds, as well as different experiences as technology users (Bouma et al. 2009; Charness 2009; Czaja and Nair 2006; Leikas 2009). They may be experienced or novices (Mayer 1997). In each case, it is essential to understand what kinds of characteristics are shared by users of a specific technical artefact. This knowledge is required when designing technologies for specific groups of people. The psychology of individual differences and personality provides information and methodologies for solving these kinds of questions (Cronbach 1984; Kline 2013).

Finally, people belong to—and co-operate in—groups of different sizes. These groups may be pairs, families, small work teams or organizations, and even nations and cultures. Social media (services allowing people to exchange text, speech, pictures, videos, and information over the Internet) is currently in a decisive position to create different groups. Irrespective of whether the group is organized around culture, work, or social pleasure, there are always common norms that are accepted and followed (Brown 2000).

Social interaction with the help of technology can involve either one-dimensional communication (interaction between the human and the machine) or two-dimensional human–human interaction (the communication is transferred from one person to another in turn). Communication can be seen as a multidimensional interaction when several people interact with each other at the same time—for example, social media applications that enable a group of persons to share each other's contexts, locations, and contextual content. Many solutions for multi-user questions and group behaviour can be found in social and organizational as well as cross-cultural psychology (Brown 2000).

A seemingly simple situation can include several major psychological aspects, for example:

- psychology of attention and perception (target discrimination);
- psychology of motor movement and movement science (handling artefact);
- working memory and LTWM (controlling ongoing interaction);
- skills acquisition and long-term memory (expertise in using);

- communication and comprehension (conveying information to machine and people);
- decision making (choosing between alternatives); and
- situation awareness and complex problem solving (being aware of what really happens).

The above-presented design and research tasks all require solid knowledge of the human mind and its operations in order to improve the user's capacity to effectively use technical systems. Good design solutions, which are in harmony with the principles of the human mind, form preconditions for the effective use of technologies. Obtaining a more concrete view of the nature of these design challenges calls the attention for systematic consideration of different aspects of psychological knowledge bases. This chapter focuses on cognitive issues.

To Perceive and to Attend

When using a technical artefact, it is necessary to receive information concerning different states of the technology and the context in which it is used. If the user is unable to acquire all the relevant information concerning its usage, an artefact will be impossible to control. For example, when loading a container onto a ship, a crane driver working on a rainy night may not see a workmate in a dark jacket and hit him with the crane. This kind of work accident can be a consequence of poor visibility and difficulty in discriminating relevant information.

Discrimination has two important stages: (1) discriminate between objects in general and (2) discriminate the relevant targets from among all the available objects. In the crane example, the challenge for the crane driver is to first notice everything that happens in the loading area, and second to discern the walking person from among other possible objects.

In psychological terms, relevant information about the state of an artefact and the task to be completed is acquired via perceptual information processes (Goldstein 2010; Rock 1983). The information is most often visual but can be auditory, tactile, somatic sensory, or even olfactory (Goldstein 2010; Proctor and Proctor 2006). The main requirement is that the user is able to find the right information at the right time and in the right location, and to use it to control the technical artefact or system. However, the user may have to perform several sub-tasks before being able to accurately discriminate the crucial information. For example, he or she has to be able to detect the target and its colour, discriminate between the target and its background, locate the target in three-dimensional space, and register its movements.

The first psychological precondition for acquiring information is thus how to find relevant information in I/O components and action contexts. The psychology of sensory and visual discrimination provides much of the key information about human performance related to these types of issues (Bruce and Tsotsos 2009; Duncan and Humphreys 1989; Neisser 1967, 1976). For example, if the sound of the given task is too similar to the background noise, the threshold of discrimination becomes too high and may lead to dangerous situations (Goldstein 2010; Proctor and Proctor 2006).

The first level of user psychological issues is information encoding by a human's sensory systems. People need to see where I/O components and contextual items are, how they are located, and how they differ from other equivalent components. These processes are essential in monitoring and controlling how technology works. One cannot interact with a technical artefact without a sensory connection to the technology and its relevant environment. However, this very basic task entails a large number of important scientific questions.

Before users can operate any I/O component in an interface, they have to be able to obtain information about the component, such as an icon on a computer display or a button in a car interface. However, perception in the scientific sense is much more complicated than simply experiencing the target I/O element. There is a wide gap between intuitive experience and scientific understanding.

In order to perceive an object, one must be able to generally discriminate it from other objects. First, there must be a physical target. This is called *the absolute discrimination threshold* (Goldstein 2010). Second, the target has to be discriminated from the background noise: *the relative discrimination threshold* (Goldstein 2010; Hecht 1924). These two aspects are present in every interaction event.

Humans perceive the environment, which is present in physical energy states. In practice, this means that individuals discriminate the information that is above the sensory thresholds and against the background (Proctor and Proctor 2006). If a factory hall has too little light, it makes it difficult for employees to discriminate the relevant I/O and other objects, and insufficient light may even cause industrial accidents. The absolute discrimination threshold is equally important in all senses. The human ear, for example, can discriminate sounds only between 20 and 20,000 Hz (Bregman 1994; Goldstein 2010). Similarly, all other sensory systems have their absolute limits, which constitute an important element of designing HTI processes.

For example, it is often difficult to see information clearly on mobile phone and tablet screens in strong sunlight. Discrimination thresholds change depending on the type of display, and also when the environmental light changes. Kindle e-books and iPads, for example, differ in how light affects discrimination. If we can discriminate only the target, the difference is called the 'just noticeable' difference. Interestingly, the discrimination threshold follows Weber's law, which means that the percentage of the magnitude change is constant over the changes of a stimulus. However, from a practical point of view, sensory thresholds are limit values, and thus can seldom be a design goal in themselves. They express minimal conditions for performance; it is almost always better to design solutions that clearly surpass minimal thresholds.

Properly designed lighting and contrast improve perceptivity. Backlights, efficient working lights, and proper highlighting can improve performance, and should thus be considered when designing working and living environments (Proctor and Proctor 2006). The safety and reliability of human performance, for example, often depends on how well illumination is designed. It is also well known that improved illumination in an office increases production (Barnaby 1980), and helps people with low vision cope better with many everyday affairs.

When illumination and other issues related to the discrimination threshold have been solved, the next issue to consider is *object perception*. In HTI processes, it is vital to correctly locate objects in space. One should also consider environmental objects as controls or displays. This requires the correct perception of objects, as well as their movements and depth; it can be called *the problem of three dimensionality* (Bruce and Green 1992) and it has two main variants. First, it is essential that users can see the location of objects in their physical surroundings in real space. Second, if using a computer screen, the user must be able to perceive a three-dimensional presentation of the object on the screen.

There are also contextual factors to be considered. The perceiving person can move quickly or be placed above the surface of the earth. Pilots, for example, must be able to correctly locate the take-off and landing zone as well as the height of the plane (Gibson 1950, 1979). Similarly, crane operators must be able to move containers into the correct places in the narrow cargo hold of a ship. This is possible only if they can see the objects and their movements in space. The question concerns the ability to place objects in a spatial arrangement as well as the movement of the perceiving person, because the perceiver's movements continuously change the retinal picture and affect the way a person perceives the environment. Three dimensionality and object perception thus form a complex set of problems for the psychology of perception (Bruce and Green 1992; Rock 1983).

The three dimensionality of a perceptual space is partly constructed on the grounds of asymmetry between retinal pictures, and partly due to distance cues (i.e., the properties of the environment, which the human mind can use to infer the distances between different objects). A typical example of this is gradient, which means that the surface has similar structures that give smaller and smaller retinal images (Fig. 4.2).

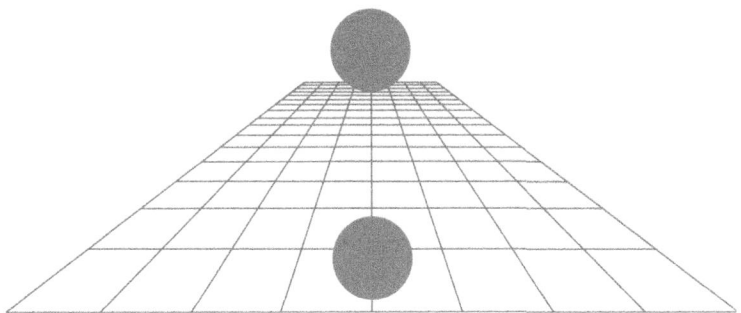

Fig. 4.2   Ponzo illusion

In this figure, the target circles are of equal size, but the upper circle seems to be larger. The gradient of the railway track suggests that the upper circle is further away, and as the retinal size of the targets is the same, it suggests that the upper circle must be larger (Bruce and Green 1992; Goldstein 2010).

This type of visual illusion can be used to construct three-dimensional spaces on two-dimensional surfaces, such as displays or paintings. Three dimensionality is also often an important problem in constructing games or other three-dimensional virtual realities. The psychology of depth and movement perception is vital in solving these design issues (Bruce and Green 1992).

The perception of movements is also based on perceptual cues. The edges of objects, for example, for a time cover a background that causes the image of a moving object (Ullman 1996). As mentioned above, an additional element in the perception of movement comes from the movements of the observer (Gibson 1950, 1979). Constructed movements are often important in improving displays. For example, when constructing virtual reality, it is often important to create body movements to avoid contradictory sensory information between the eyes and body. Without logical coordination between different sensory modalities, it is impossible to create, for example, a genuine flight simulator.

Perceptual processes in interaction include colours as well as three-dimensional moving objects (Goldstein 2010; Mollon 1982). Colours can be used for several purposes in organizing HTI processes, for example, to discriminate items, camouflage certain objects, or improve emotional usability. In HTI, colours are important in discriminating critical information. Warning signals and alarm buttons are typically discriminated by colour, and it is also important in many semiotic presentations (Tufte 1990, 1997), for example, red is the symbol for the highest temperature and blue for the lowest. Colours are also important in creating emotional dimensions for texts and other products (Desmet et al. 2001; Liu et al. 2003).

Thus, colour perception is an additional issue in encoding I/O components. While a normal human can distinguish between over a million different colours (Boyton 1988; Mollon 1982), not all people can discriminate colour information. Colour blindness can be genetic or acquired, and is much more common in men than women. There are also ethnic differences so that Eskimos have more names for the types of snows than Southern nations.

It is possible to find similar types of problems in other modalities, as discussed above about vision. Sensory information is not only visual, it may also be auditory, tactile, or vestibular (Bregman 1994; Goldstein 2010). Senses connect people with the immediate environment and provide information about the current state of their body in space or information about its current internal states.

The problem of discriminating relevant information cannot be solved by using concepts derived from perceptual information processes. The question of relevant information does not concern only the experience of perceptual space, it also requires discriminating the relevant piece of information from all other available information. However, to effectively use technology it is essential to discriminate important action-relevant information from the less important background. This process has its own principles that are studied in the psychology of human attention (Broadbent 1958; Covan 1999, 2000; Egeth and Yantis 1997; Kahnemann 1973; Pashler 1998; Styles 1997).

In emergency situations, as in the Three Mile Island nuclear accident or in industrial process breakdowns, it is often critical to extract critical information quickly so that users can respond swiftly to the demands of the situation. This may be an alarm light, a vibration, or a sound. The psychological laws of human attention help solve the design issues of discriminating between relevant and irrelevant information (Broadbent 1958, 1975; Kahnemann 1973; Chun et al. 2011; Pashler et al. 2001; Styles 1997). This can also be called the *difference between figure and ground.*

Thus, perception offers two interpretations of the same retinal image. On the one hand a duck is visible, and on the other hand a rabbit. When the duck is seen, the rabbit disappears, and vice versa. These kinds of ambiguous pictures illustrate how perception and attention are selective and distinguish between the foreground and the background (Luckiesh 1965; Styles 1997).

Perception is selective in the sense that it is often necessary to find a crucial piece of information in the environment. For example, to go out of a room, one must find the door and its handle in order to grasp it and push the door open. All these actions presuppose selective perception, and the ability to differentiate the background from the foreground. Often, background information can be distracting, for example, using a mobile phone while driving (Nasar et al. 2008).

The automotive industry is developing HUD indicators, which would project safety information onto an image of the road. The driver could see, for example, an indicator of a danger in his line of vision on the image of the road. The problem with these displays is that the driver can pay attention either to this indicator or to the road; this continuous adjustment of focus may cause an accident if the display is not well designed.

Several experimental settings have been used to investigate human attention. Visual search and cocktail party effects are perhaps the most significant ones (Broadbent 1958; Egeth and Yantis 1997; Kahnemann 1973; Neisser 1963, 1967; Pashler 1998; Treisman and Gelade 1980). In visual search tasks, target objects such as letters or numbers are searched for amongst other objects. They together form the figure 'x' whilst the rest of the letters or numbers serve as the background. In this task, people are asked to report as quickly as possible whether they can perceive an 'x' on the display. When a large number of displays are presented, it is possible to measure the average time that it takes for subjects in different conditions and contexts to find the target (Sanders and Donk 1996).

Discriminating targets from their background is important in human–technology design, because targets often entail vital information; thus, a failure to discriminate them may lead to misinterpreting situations. In radar, for example, it is absolutely vital to discriminate between an enemy aircraft and background noise. For this reason, many radars have software to make discrimination easier.

It is also well known in the theory of human attention that colours make discrimination easier when there are not too many of them (Treisman and Gelade 1980). The ability to effortlessly and quickly discriminate is called the 'pop-up' phenomenon, which is characterized by discriminative cues that are easy to pick up and thus helpful in discriminating target information. Similarly, forest workers and hunters must be discriminated so that other hunters do not accidentally shoot them. The pop-up phenomenon works well in this case too:

The visibility of the surface material of forest workers clothing was compared in situations with different lighting. These studies discovered that phosphorized yellow was the easiest to perceive for peripheral vision. Although white formed the strongest contrast, yellow was the easiest to discern against the background formed by the green forest (Isler et al. 1997).

Discrimination of the target is associated with a large set of problems, and it is one of the most fundamental tasks that users must carry out when using technology. For example, when driving a car, the driver has to detect information about the control system and signs and symbols on the windscreens of the car, other cars driving in the opposite or same lane, traffic signs, pedestrians, crossroads, buildings, trees, and lakes. In nearly all situations of normal use, the user has to discriminate objects from the background. Discrimination can also be used in reverse by hiding something in the background. For example, hiding the on/off button on many computers used to be important in preventing unauthorized usage, but this is seldom necessary.

Attention is not only a visual phenomenon. For example, people in a noisy restaurant can hear numerous auditory messages, but they can still discriminate between relevant and irrelevant messages to follow the discussion they are most interested in (Broadbent 1958; Neisser 1967; Pashler 1998; Styles 1997).

Evacuation planning includes the design of audible alarms that must be heard against a background of noise. Therefore, alarm designers know that they should adjust the frequency of alarms to a level that is clearly different from that of the background noise (Häkkinen 2010). Manipulation of volume is not desirable because the human auditory system is more capable of discriminating frequency than intensity, and the auditory attention benefits from differences in frequency.

It is also important to remember that the object of attention varies. An individual's gaze is targeted at the most significant items for current action, and it lingers the longest on essential information. For this reason, eye movements have become an important method of studying target perception (Baccino and Manunta 2005; Henderson and Hollingworth 1999; Josephson and Holmes 2002; Mackworth and Morandi 1967;

Yarbus 1967). For example, the design of panels in industrial process or aviation control rooms must follow this principle. The most relevant and important displays must be located within the attentional focus of the operator (Norman 2002). Consequently, when designing a user interface, the factors that direct attention must be taken into account.

Both follow-up of the object of attention and eye movements have an important role in usability research. For example, Josephson and Holmes (2002) studied the scan paths that people's eyes follow when they read information on web pages, and Howarth (1999) investigated how the eye adapts to the distance from a stimulus created in virtual reality, and found that adaptation was different in virtual reality than in real-life situations.

Unintentional direction of attention can be considered the basic form of directing attention (Chun et al. 2011; Eimer et al. 1996; Styles 1997). It is a fundamental mechanism for reacting to sudden changes, such as a threat, in a person's environment (Eimer et al. 1996; van der Heijden 1992, 1996). A rapid shift in unintentional attention is also called *orientation*. It signifies a comprehensive reaction, such as an attack or an escape. These kinds of reactions also include, for instance, a sudden increase in adrenaline secretion and nervous reactions (Eimer et al. 1996; Posner 1980). When people direct their attention to an object based on their prior intention, this is called *self-controlled* or *voluntary attention* (Chun et al. 2011; Eimer et al. 1996; Styles 1997) because this type of a shift in attention originates from the human's own will and need.

The laws of target discrimination in orientation have been used to improve the efficiency of the message. For example, pop-up icons, which interrupt the continuous state of working, call for human attention by causing an orientation reaction (Posner 1980). Yet since this also breaks the normal course of the main action (Shneiderman and Plaisant 2005), many people use pop-up blockers. On the other hand, there are many situations, such as when a program crashes, when it is vital that the message is well discriminated even during severe stress.

When people repeat the same perceptual-motor task in the same circumstances, they start to improve their performance. It becomes faster, less error prone, unconscious, and independent of will until finally it requires few cognitive resources (Neumann 1984; Schneider et al. 1984). This learning process is called *automatization* (James 1890; Schneider et al. 1984), which has received major attention. Automatization is very important when learning to use technologies. Automatic processing can be explained in terms of feature discrimination. People learn the systems of discriminative cues, which enable them to process information faster. They need not put any effort into discriminating a target object, as their information-processing system easily picks up the decisive information.

For several reasons, it is good that human performance is automatized in many HTI processes. It makes performance fast, error-free, not capacity demanding and therefore protected against such sudden factors as panic (Neumann 1984). The only problem is that in an unexpected situation, automatization may cause people to make serious errors, as they cannot adapt effectively enough to the changing demands of the advanced technologies.

People can have different roles related to controlling artefacts. For example, they monitor the performance of artefacts in many different ways. Normally, people are continuously monitoring the performance of the artefacts. They seek information and respond to it immediately, such as when driving a car on a curvy road; they are 'in the loop'. However, the more control that is transferred to the artefacts, the less involved people are in the performance of the technology. Good examples of this are provided by autonomous systems. In a problem situation, the interaction process with complex technology becomes increasingly demanding. This phenomenon is called being 'out of the loop' (Endsley and Jones 2011).

An additional problem linked to attention in interaction processes is the amount of time that people are able to efficiently and reliably observe the same target. Many professions require constant vigilance. People may work to observe, for instance, radar or a monitor for a long time, without any deviant behaviour being shown on the display. This kind of continuous observation task may eventually weaken one's attention or lower the level of vigilance, which can be hazardous in many critical surveillance tasks. These are common challenges in interaction design, for example, in the process industry (Koelega 1996). Especially in fields in which even small breaks (not to mention longer delays in manufacturing processes) become very expensive or risky in some way, the work tasks should be properly designed and examined at intervals. It is quite typical that emergency processes, for instance, have not been tested for years, and then do not work properly when they are suddenly needed.

If the level of attention lapses due to lower levels of vigilance, the performance level decreases. When this happens, the probability of failure rises and the number of errors increases. Although these effects are to some extent individual, problems related to vigilance can be taken into account in technology design (Koelega 1996). For example, tasks that require continuous attention should not last long, and should be divided into periods with breaks. Vigilance begins to weaken after half an hour of continuous attention and a continuous work performance of two hours can be too demanding. People should also be trained efficiently in order to minimize the number of processes one has to keep in mind at the same time.

Fatigue of night shift workers is a typical vigilance-related problem. Bonnefond et al. (2001) conducted a one-year study of workers who were allowed to sleep for an hour during a night shift and found that this arrangement improved vigilance. This research suggests that this type of arrangement is useful where the work task requires a person's full attention, and if it is possible to arrange a sleeping break at the work place. The study itself is a good example of usability design analysing vigilance.

Another important concept of perceptual psychology is called affordance (Gibson 1979), which refers to how a visual object is perceived from the point of view of action. For example, a glass is perceived differently when taking it from a cupboard, filling it, and drinking from it as opposed to throwing it at a referee in a football match (Gibson 1979; Gielo-Perczak and Karwowski 2003; Rasmussen 1983). Gibson (1979) and colleagues note that affordance also entails information about the state of the object that is related to what people *do*. Because of this close relationship between affordance and action, it is possible to think of affordances as attention phenomena.

If there are several alternative courses for actions, affordance allows individuals to attend to objects from the correct point of view (Gibson 1979). Thus, their eyes, hands, limbs, and body anticipate the next moment of the action depending on its affordance. When a person takes up a pen to sign a document, he or she does this in a different way than if they throw it away, and when they drink from a teacup it is not the same action as washing the cup. The phenomenon of affordance requires not taking perception as an absolute but rather linking it to the nature of a current action. This concept has been introduced into HCI research to develop its ecological relevance (Forrester and Reason 1990).

The first sub-task in interaction design for usability, *discriminating relevant information*, has a large number of links with the basic psychology of perception and attention. While knowledge of human perception and attention can be used to define and solve important interaction problems, the main problem is to analyse the technological interaction tasks effectively and to consider the coherence between them and the tradition of attentional psychology.

To Respond and to Act

Senses have a significant role in collecting information about different states of technology. Some kinds of tools (e.g., hands or feet) are needed in order to respond to this information and to control the technical artefact. Human psychomotor action plays a central role in using technology. Action itself is a hierarchical process with several different levels (moving a finger, moving an arm), but the control level consists of the thoughts about what a person is doing and why. Each of these levels makes it possible to understand some aspects of the action in question.

Within a specific context, *the ultimate action goal* serves as a person's highest level of action and as a trigger in using technologies. A ship is built to transport cargo and to ferry passengers from one port to another. Having this particular everyday action (moving a ship to a new port) as the ultimate action goal explains why people use ships in general. In addition to this overall task, each individual act needed in steering the ship has its own goal and intentional content (Jeannerod 2007). This thinking is supported by Searle (1993), who distinguishes between prior intentions and intentions in action (Jeannerod 2007) and effectively illustrates the distinction between action in the life of users and action in using a technical artefact or system. This chapter mainly focuses on the *action in using* a technology.

Human psychomotor movements are present in every human action and have a necessary role in interacting with technologies (Bridger 2009;

Karwowski 2006). For example, in addition to the movements of hands, feet, and eyes, speech is produced by the activity of the psychomotor system associated with the larynx and the lungs (Rayner 1998).

Perception has a direct connection to motor functions (Olson and Olson 2003; Schmidt 1975; Schmidt and Lee 2011). It automatically monitors the environment whilst the person is, for example, walking. The connection of perception and attention to motor functions is termed *perceptual-motor cycle* or *sensory-motor integrations* (Fotowat and Gabbiani 2011; Schmidt 1975). This system has been equally important to Stone Age hunters and modern office workers. With the help of perceptual information processing, it is possible to perceive the surrounding environment and respond to it with the operations of the psychomotor system. Without perception our motor functions would be helpless, and without a motor system our perception would be senseless.

An important property of the human psychomotor system is its hierarchical character. People use different types of *motor movement patterns* that are partly linked; thus, each higher-level action depends on different lower-level motor patterns (Gallese and Lakoff 2005; Kawato et al. 1987). Lower-level patterns include, for example, posture, instinctive reflexes, and intentional motor functions such as stretching or pinching. These functions support a large variety of higher-level actions of complex motor patterns from walking and typing to gymnastics and dancing.

Technical environments often include restrictions that force people to work in an unergonomic manner in terms of motor functions. Using a mouse, for example, can cause elbow and shoulder problems in the long run (Murata et al. 2003). Ergonomics is thus needed in work environments in order to develop ideal posture models and ways of working with new technologies (Bridger 2009). Posture as a zero movement is an idealized concept here, as it is present in all complex movements and the realization of motor tasks depends on the success of the posture adopted.

The most fundamental motor function is a *reflex* (Pavlov 1927). For example, when someone puts his or her hand on a hot surface, the hand is withdrawn quickly and automatically without any conscious control of the motor function. The information from the finger goes to the spinal cord and from there to the muscles responsible for withdrawing. The information goes more slowly to the higher levels as the hand is already moving away. Very close to this type of reflex are many learnt automatic responses that operate unconsciously. Typing or driving a car, for example, are built on unconscious low-level motor patterns of fingers and hands (Card et al. 1983; Rumelhart and Norman 1982; Salthouse 1986).

Instinctive and automatic motor functions are important, as they are used in situations that require rapid reactions. Such functions are vital since they help avoid danger. Even today, one of their main purposes is to help humans react to immediate changes in their environment. In earlier days, these kinds of self-protection mechanisms were important in fighting off external attacks, but today they are helpful, for example, in avoiding traffic accidents.

The second type of motor function is formed by a large set of simple *intentional motor functions*. When a hand reaches out towards a control device, the individual is intentionally leading it towards a goal. Whereas automatic motor functions are subconscious and unintentional, intentional motor functions—although based on evolutionarily developed patterns—essentially rely on higher cognitive processes. Various standard functions such as walking, reaching, or grasping are examples of these action patterns in humans, and therefore they also hold a central position in usability design. People employ these functions when they use technologies such as a mouse or keyboard (Lewis et al. 1997).

Input modes for user interfaces are becoming increasingly advanced with the development of ubiquitous, ambient, and pervasive technology. For example, many modern artefacts from TVs to museum guides are already based on gesture recognition. In addition to recognizing the intentional motor functions of users, a great deal of effort is being put into developing smart environments, which respond automatically to the behaviour of the user. That is to say, to the unintentional motor functions of the user.

Finally, complex schematic motor functions also need to be considered (Schmidt 1975; Schmidt and Lee 2011). In addition to simple and basically automatic procedures, human beings also execute complex and learnt *series of motor functions*. Such series can be recalled from memory and applied in an adjustable manner depending on the current situation. In technology usage, to control complex machines generally requires motor actions of the highest level, and it usually takes a long time to learn them. Handicraft provides a good example of complicated motor patterns. For example, the actions of a goldsmith or watchmaker require dexterity and hand–eye coordination.

High precision and dexterity are fundamental prerequisites of the work. In fact, surprisingly little is known about controlling precision and dexterity, although this subject is constantly under discussion, for example, in the expanding area of robotics research. It is difficult to develop and program robots to perform human-like motor functions with adaptability, accuracy, and versatility, and the unconscious nature of motion control further complicates the issue. In addition to robotics, different kinds of simulators have found their way into the research and development of technology and human psychomotor action. In surgery, for example, where human motor functions are a necessary complement to a high level of cognitive expertise, simulators have become increasingly popular training tools. They make it possible to practice risky operations without putting patients in danger.

Elementary motor patterns can be either automatic or intentional. They form basic elements of studying the nature of complex series of motor functions (Schmidt and Lee 2011). Such patterns include, for example, grasping a drinking glass, lifting weights, and shaking hands. Variations of these patterns are unconsciously repeated (Desmurget et al. 1999; Schmidt and Lee 2011). Indeed, all complex motor functions are variations and combinations of elementary motor patterns.

Basic motor patterns may change to some extent as a result of learning. The actions of a professional cricket batsman provide an interesting example. In this game, the ball is bowled to a batsman, who has approximately half a second to adjust his strike to the movement of the ball. Land and McLeod (2000) studied the eye movements of cricket players at different levels using an eye-movement camera. They discovered variations in the ability to follow the movement of the ball between players at different levels. Every player directed his eyes in the same way to the initial trajectory and ground contact point of the ball, but the speed of following the ball with one's eyes clearly distinguished players at different levels. Years of practice had made the eye movements of the more experienced players quicker and more appropriate than those with less experience. Similarly, it is possible to train dexterity and other motor functions required in the usage of different technical artefacts.

Motor patterns are holistic in nature to the extent that a change in one motor function creates a change in the whole motor pattern. In this sense, user interface paradigms should be developed following the holistic patterns of users, or at least they should not require severe changes in natural motor patterns. For example, when operating smartphones whilst walking, users are forced to acquire new types of motor patterns, that is, to watch the behaviour of the artefact instead of observing the surrounding environment. This change in natural motor patterns increases the risk of tripping or falling. Therefore, it is vital to analyse in detail the possible changes in motor patterns and postures in the usage of a developed artefact.

As a whole, human psychomotor action highlights a large set of practical problems that have to be considered when designing HTI processes and user interfaces. These problems are mostly investigated in the area of ergonomics and human factors research (Bridger 2009; Karwowski 2006), but understanding the basic principles is essential in any design target. Human psychomotor action forms an important conceptual basis for user interface design. Its characteristics determine the structure and properties of the input devices and controls of technology.

To Learn and to Remember

The technologies that human beings have created are far beyond the capacity of any other animals. People's ability to create and construct is mainly a result of the human capacity to keep large associative structures in the memory and, consequently, to learn large systems of knowledge such as languages. Memory and learning are central factors in understanding HTI; without them, people would be unable to create technologies to serve them. These two processes are always present when people use technologies, from learning user guides to highly educated professional use of technologies. Computer programmers, for example, must keep in mind aspects of programs such as reserved words, function names, typical algorithms, the way they are used to construct well-formed expressions, and the structure of the program they are working with at that moment. They have to remember large amounts of symbolic information to be able to carry out programming tasks. In order to understand this kind of complex activity, it is essential to understand how people have learnt to program—especially how their memory (which makes the learning possible) operates.

The importance of memory and learning has been long understood in the field of interaction. Memorability and learnability are acknowledged usability design criteria, and one goal of usability testing has been to illustrate deficits in these areas (Nielsen 1993). One way to do this is to measure values in remembering and learning compared to other interaction processes. A completely different and more challenging task is to find out what kind of changes should be made to really improve users' abilities to remember relevant and required information. The improvements presuppose understanding the underlying memory processes and how they can be influenced. For this reason, usability design and research should be intimately connected with the user psychology of the human memory (Saariluoma and Oulasvirta 2010).

Memory is as basic a cognitive process as perception and attention. It is needed in motor functions, and it forms the basis of human skills. The basic memory processes, as seen in memory research, form complex networks of questions that relate to how people encode, store, and retrieve information in their minds and what kinds of properties the many important memory systems have (Baddeley 1990, 2007). The importance of memory has been understood within the usability community from the very beginning (Nielsen 1993). The present usability discourse only addresses the concepts of memory on a relatively basic level, and detailed analysis of the connections between HTI and memory processes is still in a relatively early state. There is much work to be done to illustrate what aspects of the human memory are important, and in what ways, when developing HTI solutions.

An example of such research is the connection between pictorial recognition memory and GUIs. The superiority of GUIs compared to symbolic interfaces has had an enormous impact on interaction studies. Not only smartphones are touchscreens; many other instruments have been redesigned. For example, dentists today use touchscreens to control their instruments. The breakthrough of two revolutionary technologies—GUIs and touchscreen technologies—has been based on the use of pictorial recognition memory (Shepard 1967), which is a more effective form of learning than text-based solutions.

Psychologists have known for at least 30 years that human pictorial memory and the capacity to learn to recognize pictures is much better than the human capacity to remember and recognize symbolic strings (Standing 1973; Standing et al. 1970). Using pictorial information as the basic interaction model makes it much easier for users to interact with technology compared to symbolic interaction paradigms. Therefore, it is difficult to understand why this fact was overlooked for so long: many usability tests illustrated the importance of pictorial information long before the development of touchscreens (Ramsay and Nielsen 2000).

One of the most prominent questions in the research on human memory has concerned different types of memory. Human memory is divided into several types of storage; the most important in this context are working memory, LTWM, and long-term memory. These systems have different roles and explain different interaction-relevant phenomena in human information processing, which is responsible for controlling human actions. To understand the importance of these explanatory phenomena of memory, it is essential to consider each of them separately and to clarify their role in HTI processes. The overall structure of human memory is presented in Fig. 4.3.

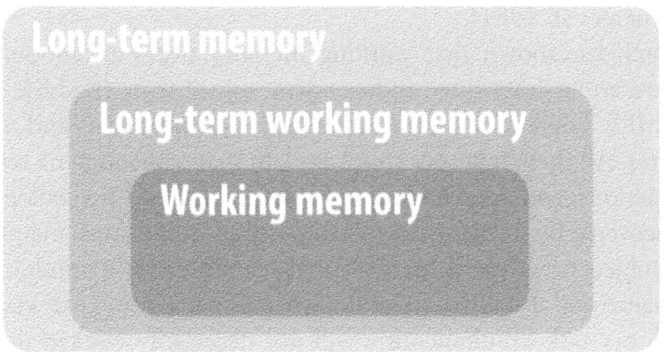

Fig. 4.3 The overall structure of human memory

Intuitively, people tend to think that remembering concerns only past experiences. Memory researchers, however, argue that remembering also includes keeping specific information active. Smooth human performance presupposes the ability to anticipate the demands of the tasks that follow. In order to prepare for these tasks, the human information-processing (working memory) system must keep information in the memory in an active state (Baddeley 1986, 1990, 2007, 2012; Ericsson 2006; Ericsson and Kintsch 1995). The systems of preparation can be seen in subconsciously activated associative networks detected in priming tasks (Anderson 1976, 1983, 1993). Scientific experiments have demonstrated that activating one piece of information, such as seeing the word 'nurse', leads to the activation of related information, such as the word 'doctor', but not to the activation of unrelated information, such as 'butter'. This very systematic phenomenon illustrates the human mind's capability to better prepare for the emergence of related than unrelated issues.

The system of memories is practical, as it enables people to distinguish between task-related and relevant information during interaction processes. For example, different kinds of warning systems can improve users' performance in critical situations (Häkkinen 2010), but they may make users neglect design mistakes. Also, a programmer who must take care of thousands of lines of code whilst focusing on writing a single object can use his working memory to keep the immediately needed information about the object active. At the same time, he keeps the program as a whole in LTWM, and the rest of the knowledge in long-term memory (McKeithen et al. 1981).

Research has shown that working memory stores are modular, and that there are at least three independent sub-systems (Baddeley 1986, 2007, 2012; Logie 1995): central executive, visuospatial, and articulatory stores. All these memory stores explain many phenomena related to human actions, and are therefore fundamental for user psychology. They were discovered by studying how simultaneous tasks disturb memory. If a human subject tries to perform two visual or auditory tasks simultaneously, performance will usually deteriorate, whereas performing one visual and one auditory task simultaneously will not necessarily weaken the parallel performance (Baddeley 1986, 2007, 2012; Logie 1995). In practice, these systems are in intimate collaboration, but depending on the properties of a particular sub-task, they may have a different value in solving the particular problem.

Miller (1956) discovered one of the main properties of the working memory when he examined people's performance in various tasks that required memory, such as multidimensional evaluations. Based on his analysis of the test results, Miller discovered that human immediate memory is very limited in capacity. Many others after him discovered the same thing, and the number of empirical observations began to increase. Based on these results, a division between working memory and long-term memory was established (e.g., Atkinson and Shiffrin 1968; Miller 1956). It is now known that working memory is temporary in nature, and the information it contains easily decays when new information enters.

The storage capacity of the working memory has been estimated using various methods (Broadbent 1975; Covan 1999, 2000; Crowder 1976; Gregg 1986; Miller 1956). These estimates have concluded that the scope of the working memory is around four to seven separate units. This capacity is so small that it causes problems in all tasks that load the working memory with unrelated and new symbolic information as well as new systems of symbols; it also makes learning cognitive skills very time consuming (Newell and Rosebloom 1981). Thus, programming is a good example of the necessity of processing complex symbols (Anderson 1993; McKeithen et al. 1981).

Event sequences in controlling technologies often require accurate memory for relatively long chains of performance. Even such a simple task as adjusting the time on an electronic clock may require long sequences of keystrokes. Learning them may be time consuming, because the system does not necessarily provide intuitive and logical support for comprehending what should be the next sub-task to reach the expected goal state of the artefact.

In scientific experiments, the capacity of the short-term working memory has proved to be rather small. This creates a new dilemma: how is it possible for a creature with this little capacity to manage large and complex devices such as aeroplanes and rockets? The answer is in people's ability to code individual bits into larger entities: chunks and memory units (Fig. 4.4). Working memory limits the number of units, but not their size. An individual letter, word, or sentence can all comprise a nearly

my cat likes milk in the mornings

soy ltigr tmka chnim ilesn nkiem

Fig. 4.4 The first sentence is easier to remember although the sentences comprise the same letters

equal load for the human memory, although they are unequal in the sense that a sentence contains numerous letters (Baddeley 1990; Crowder 1976).

Human cognition can circumvent the limits of the working memory (Miller 1956). Programmers, for example, can keep in mind long pieces of code by organizing it into sense-making wholes. They can perceive in the program text familiar 'chunks' such as loops, structures, search algorithms, and sorting methods. They also have an idea of the structure of the whole piece of code they are working with, and can thus effectively utilize chunking mechanisms. Structure and object-oriented programming paradigms particularly help support information encoding during programming.

The limited capacity of the human working memory is an essential component in analysing and explaining cognitive workload (Wickens and Holands 2000). Researchers have used a series of several equivalent tests to prove that humans store extensive information networks in their memory. According to the type of stored information, these chunks can be called motor, visual, or other schemas, scripts, or semantic networks (Markman 1999). Different elements of these memory structures become associated with each other and form vast networks, which can operate as chunks.

Since chunking may be based on semantic associations between items, it is easier to remember words that can be naturally categorized (Bower 1970; Bower et al. 1969), for example, into categories such as WEAPONS (rifle, sword, dagger, pike) and FRUITS (apple, pear, orange, etc.) than those that are totally unrelated to each other. Moreover, items' visual and spatial properties may also enable chunking. Thus, unrelated words that can be chunked into bizarre images may aid remembering. For example, words like hat, whale, cigarette, spectacles, and an overcoat can be more easily recalled when imagined as a bespectacled whale smoking a cigarette and wearing a hat and an overcoat. Likewise, when working with spreadsheets, people may find it easier to form implicit images (Saariluoma and Sajaniemi 1989, 1994), and icons can be grouped semantically and spatially.

Although working memory limitations can cause problems in any HTI task, it is often possible to support users' performance by external memory cues—that is, by visual information indicating the state of the user action. 'Breadcrumbs' are typical examples of navigation tools that have been developed to help to keep relevant information active in the minds of users.

Human working memory performance is weakened considerably in secondary tasks, when two actions are performed at the same time. Even tasks that are independent of modality, such as the production of random numbers, may decrease the performance level. This has been thought to disturb the division of tasks between the stores of working memory (Baddeley 1986, 2007, 2012). The visual sub-system is harmed by visual secondary tasks and the auditory system by auditory tasks. A typical example of a visual dual task is in-vehicle information systems such as navigators. A driver's main task is to control the car, but he or she also needs information about the route; he or she will be at risk if they take their focal vision away from the road to look at the navigator.

There is yet another type of memory store that has been found. Many experiments have shown that experts can overcome the limited capacity of the working memory by retaining information relevant to their domain of specialization while undertaking secondary tasks (Ericsson and Kintsch 1995). While counting backwards from a three-digit number will considerably disturb remembering trigrams (sequences of three letters such as BBC or MEP), it is of little consequence to skilled chess players who are used to storing chess positions in their memory (Charness 1976; Saariluoma 1995). Similar results have been obtained in studies on experts' memories in various fields (Saariluoma 1995), which indicate that skill-related information lies in a store that is deeper than the working memory: LTWM (Ericsson and Kintsch 1995). Blindfold chess players can, for example, learn in an hour the locations of thousands of pieces in numerous games (Saariluoma 1997). This capacity is based on pre-learned patterns of pieces that operate as chunks in the working memory. LTWM stores large amounts of information in a more permanent state than the actual working memory in order to protect details required by the active task at hand from incidental distractions. For instance, experienced taxi drivers can store even the longest route maps in this memory (Kalakoski and Saariluoma 2001). Likewise, although programmers see only a few lines of code at a time, programs can contain thousands of lines and have several layers (McKeithen et al. 1981). Most programming paradigms, such as structured or object-oriented programming, entail ways to facilitate associative structures and command program codes. From the machine's point of view, the choice of the programming paradigm is irrelevant. Using corresponding structures that are efficient and conducive to memory is also advisable when developing other interaction environments.

LTWM is a good storage for any domain-specific information, as it protects information during interruptions and makes it possible to keep in mind new information even though it was not the object of attention. This is called episodic indexing. It is a typical instance of an intriguing phenomenon related to long-term memory theory. It is relevant, for instance, in research on interruptions when using technologies (Oulasvirta and Saariluoma 2004, 2006). A person who is making coffee can be disturbed by a telephone call. After the call, she should return to making the coffee, but people with memory disorders do not remember to do so.

Memory is a necessary precondition for *learning* (Bower 1981). Learning enables people to accumulate their experiences and, as a consequence, change their behaviour and act more appropriately in new situations. Everything in a person's memory must be somehow learned; no one knows inherently how technical artefacts work. New information is stored in the long-term memory, thus learning increases the amount of information stored there.

Modern theories of cognitive learning (e.g., Anderson 1993) consider that the learning cycle begins with the presentation of information, usually in a verbal form, which describes how to perform a task. This information is stored in a declarative way with the help of training, and it becomes procedural knowledge. Analyses of these observations indicate that users do not intensively try to learn a whole system, but instead only learn the most urgent work-related information. After learning this information, they start an exploratory process of trial and error. This type of learning has been called 'exploratory' and has become one of the most investigated in *cognitive ergonomics* (Bridger 2009).

A feature that differentiates exploratory learning from other types of learning is that in exploratory learning, the trainee performs a search process that does not occur in other types of learning. When students learn physics, they are provided with examples that explain the concepts to be learned. When they learn through exploration, they look at the possible interface objects (menus, icons, etc.) that may be related to the task they want to perform. The search is performed with a task in mind in most cases, although it is common to find users who explore the interface with no particular goal in order to find out what they should do. In any case, the most important aspect to consider if a designer wants to facilitate the search (and therefore encourage exploratory learning) is to design the interface so that objects on the screen can be easily related to the task he or she wants to learn to do.

The most common learning strategy is a search strategy called 'tracking tags', which involves the user searching the user interface and trying to find a word, phrase, icon, or any other element that is connected with the task he wants to perform. For example, if a user of a word processor wants to know how to type text in bold, he would search on the screen for the word 'bold'. If this is not found, he would try other words, such as 'source', 'format', and so forth. Novice users typically try to search for common words that have a meaning in different contexts but not necessarily in word processor usage. Those who are more proficient in using a word processor can guess that to select a bold font, one should look for 'Format font'.

Users can also rely on such general strategies to solve search problems, and open a menu and fully explore it before opening another menu (Rieman et al. 1996). However, the aspect that has been highlighted by scanning learning is that the search is performed based on the semantic similarity between the task the user wants to perform and the label on the user interface. Soto (1999) has shown that the time needed to learn to perform a task on a new interface depends on the semantic relationship between the task and the label. The semantic relationship was measured empirically using the Latent Semantic Analysis (LSA) (Landauer and Dumais 1997).

Another important learning phenomenon in the context of HTI is *transfer* (Helfenstein and Saariluoma 2006). The transfer of learning occurs when learning a task facilitates the learning of another, similar task. Two views have been proposed to explain the effect of transfer. According to the first, originally proposed by Thorndike and Woodworth (1901), transfer involves similar or identical elements in the two learning situations. The second view is known as transfer through principles. It was originally proposed by Judd (1902), who suggested that transfer does not occur when there are common elements in the learning situations, but when principles or rules learned in one situation may be applied to a new situation (Singley and Anderson 1987). In this light, style guidelines are important because they keep the interaction culture sufficiently similar and the design solutions somewhat consistent, which makes it much easier for users to learn and use different applications.

To Comprehend

HTI is ultimately communication by nature. It is based on an activity in which the user feeds signs to the artefact, to which the artefact responds in an adequate way. In information technology, the meaning of signs is obvious, as the artefacts are all symbol-processing machines. While this is also the case with traditional electromechanical machines, these systems of communicative signs are different, as the information exchange is based on natural signs. In what follows, this central role of signs in HTI will be examined through semiotic concepts. The way to organize I/O actions is to develop reasonable communication models and languages for information exchange between people and artefacts. Systems of signs—and the respective distinctions between the expected machine operations—make it possible for people to use and control technologies, but without understanding the general principles of semiotic systems and semiosis, it is hard to design human–artefact communication.

Semiotic codes, such as natural language, are systems of arbitrary symbols connected to underlying concepts or references. One could also say that they are thoughts. Via thoughts, signs are linked with actual referents (Ogden and Richards 1923; Lyons 1977). Thus, the word 'dog' is connected to the concept of dog (the reference) one has in mind, and finally to the actual dog (the referent). The meaning of the word depends on the way one thinks about the referent. An atom today, for example, is not perceived in the way it was in Dalton's or Democritus' time. One branch of semiotics, called engineering semiotics, is designed to clarify the code systems used in HTI (de Souza 2005); in general, understanding the foundations of semiotic systems presupposes an analysis of the mind.

People use symbol systems to communicate with each other in different ways. Such symbols evoke conceptual representations in peoples' minds. These symbolic systems—and in a wider sense, any semiotic system (i.e., the possibility of using signs)—also enable people to communicate with machines. In order to analyse how people really use technical signs during interaction, one can apply technological psycho-semiotic thinking (Saariluoma and Rousi 2015).

As technology has become more complex, people have learned to use and construct complex semiotic systems to communicate with machines and devices. The difference between human-to-human and human-to-machine communication is that machines lack the mental systems to interpret symbols. Symbols used by people evoke deterministic and causal processes, but technical artefacts are unable to interpret and use symbols in a similar, flexible, redundant, and creative manner (Chandler 2007; Eco 1976; Lyons 1977). This is why human–machine communication is said to be mechanistic. It does share many properties typical of human-to-human communication, which in turn makes it easier for people to use machines.

A common solution to human–machine communication is based on artificial pseudo-languages such as LISP. This kind of communication is intentional and direct. The basic idea in constructing such artificial languages is that they make it easier for people to remember how to operate complex machines compared to only using the signs of zero and one (Barnard and Grudin 1988; Moran 1981). The words of technical pseudo-languages provide memory cues and support, and allow a huge set of possible instructions for technical artefacts given by people.

However, command languages are not entirely unproblematic. Their commands may have misleading contents, or they may be difficult to discriminate. For example, error diagnostics and supervisory languages must be made consistent and semantically rational to allow the fast and intuitive use of complex machines. When designing command languages, it is necessary to understand how the contents of the information are encoded in the minds of the users. The language is meant to deliver users' commands to machines as well as information about the state of the machine to users. For example, it is possible to use eye-movement detectors to pick up which part of a screen the user is looking at, and to use this information as a pointing device. In this way eyes can, like a cursor, indicate the location of the user's intention on the screen.

Part of human communication is *non-verbal* in nature. Human paralanguage (such as body language) is a constantly used non-verbal communication system in human-to-human communication. Thus, when reading other people's emotions, faces play a constant and important role in the human mind. Similar processes are under development in modern computing; in ubiquitous computing, no direct verbal guidance is needed. Instead, machines deduce from human behaviour what the users would like to do, and users understand intuitively what they should do to reach their goals.

Sometimes, communication between people and artefacts is so elementary that it is difficult to notice that these cases are also considered human–machine communication processes. Many common-life tools are *intuitive*. They are easy to use and easy to learn to use. However, as Norman (2002) has illustrated, there are also many tools that are totally incomprehensible. Even a door handle that needs to be lifted instead of pushed down can cause substantial difficulties for users who are used to the standard usage. Intuition is thus very much a matter of tradition and culture. Non-intuitive tools are non-communicative, as they more or less turn the 'meanings' of 'words' upside down.

A common example of non-verbal communication is provided by GUIs, which do not require any specific knowledge from users. These pictorial metaphors and simple interaction models, as in the case of smartphones, follow interaction patterns that are already familiar to users. They can thus effectively exploit users' existing knowledge of the general characteristics of GUIs. The solutions have been obvious and, for example, the root-knot-based navigation solution is ingeniously used in mobile phones, for example. People can reach their goals using obvious metaphorical information without the need to inspect thick user manuals.

Icon-, Windows- and menu-based interfaces thus provide examples of ways to develop implicit communication between machines and users. The selection of the icon contents and their organization mean that interaction situations and their requirements are obvious for users. Using binoculars as the emblem for search is a good example. The effective use of transfer, as in many Windows systems, spreads information from one situation to another so that learning one way of doing things essentially improves the interpretation of other interaction events.

Visualization is also an important form of communication that is possible thanks to modern technologies. Analysis and understanding of visual programming languages and statistical visualization, virtual reality, and animation are examples of the scope and power of visualization (Brown et al. 1995). Information visualization enables designers to make complex issues more concrete (Larkin et al. 1980; Tufte 1990, 1997). Moreover, virtual reality-based techniques make it possible to communicate real-life experiences of such complex issues as architectural objects.

Visualization often serves to make symbolically visible the type of characteristics that are normally not perceivable—for example, the temperature of different parts of an engine (red for hot and blue for cold). Similarly, crisis areas on maps are often depicted in red and peaceful areas in blue. However, since visualization includes its own system of symbols and meanings, the cultural relativity of symbols should be considered when developing visualizations. Thus, visualization is a means of promoting apperception and is not, therefore, connected only to perception.

Organizing human–machine communication combines semiotic and psychological research—that is, technological psycho-semiotics (Saariluoma and Rousi 2015). This field introduces a large set of questions, as it is essential to have psychological grounds for semiotic phenomena. Such issues as distinguishing signs or their cultural contents are typically multidisciplinary issues, which presuppose an understanding of the underlying psychology. Especially relevant is the psychology of comprehension. Three fundamental aspects of the psychology of comprehension are examined below: the concepts of mental representations, mental models, and metal contents.

In command languages, navigation, and other interaction situations, not all the required information to construct a *mental representation* is visible (Markman 1999; Newell and Simon 1972). When the information contents of mental representations or *mental contents* in using technical artefacts are considered, there are numerous *mental content elements* that have no sensory or retinal correlates. When users navigate in a menu system they represent their goal, but these goals are not present as such on the screen that they see. Similarly, programmers usually see a tiny part of the whole program code, but their knowledge of other pieces of code (which may not even have been written yet) influences what they do. Contents such as 'possible', 'infinite', and 'eternal' also have no sensory equivalence (Saariluoma 1995).

Imagine that a designer is developing software for a bank. She knows that the user interface must include a set of user interface components used in banking tasks. It is not important whether these components are then represented by icons or menus, or in some other way, but an important issue remains: how the hierarchical spatial organization of the user interface will reflect the mental representations of the user. Therefore, the designer will need to analyse the representational structure of the components involved in the task. Although the designer could formally analyse the tasks that the users perform—and thereafter establish the semantic relationships between different user interface components—it is also possible to use a simple and effective procedure that involves applying indirect techniques of knowledge elicitation.

Information on the nature of mental representations of the user interface components related to banking in users' minds can be obtained with statistical methods such as pathfinder and latent semantic analysis (Landauer 2007). This procedure generates a network in which the concepts are the nodes, and the links represent the relationships between them. If there is a link between two concepts, the designer could interpret that in the user's memory there is a relationship between them. In this example, the central concept in the network of banking is 'clients'. This can be interpreted as meaning when people think of a banking task, the first component that should be available is the customer database (names, ID numbers, account numbers, etc.) and after this, the operation that the user wants to perform.

Comprehension during interaction largely depends on how users mentally represent the signs and situations. Good usage presupposes having relevant mental models of the artefact (Gentner and Stevens 1983). Interacting with a system relays knowledge of its structure, functioning, codes, and the meanings of individual signs. This knowledge is called the *mental model* in the field of design artefacts, and its assessment and measurement are considered crucial for designing a system with which a person can interact effectively (Johnson-Laird 1983, 2008). Mental models are thus representations of the world that human beings have in their minds (Cañas et al. 2001; Johnson-Laird 1983, 2008).

The term 'mental model' has been used in research on interaction with physical systems. A person creates a representation in their working memory by combining the knowledge stored in the long-term memory with information extracted from the characteristics of the task (Gentner and Stevens 1983). In such cases, the information stored in the long-term memory relevant to HTI is related to the knowledge of the structure and operation of the device.

Mental models to study the interaction with artefacts can be viewed from various perspectives. For example, a good mental model is considered important in the acquisition of skills for manual control systems. Recently, interest has expanded to include monitoring tasks of automatically controlled systems in which the skills that come into play are detection, diagnosis, and fault compensation (Rasmussen and Rouse 1981). In the study areas of manual control and monitoring, mental models are used to perform calculations on the expected control performance when interacting with a system.

It has often been demonstrated that when individuals interact with a computer, they acquire knowledge about its structure and functioning. Research has shown, for example, that acquiring an adequate mental model of the computer makes learning a programming language easier (Moran 1981; Navarro-Prieto and Canas 2001).

Thus, comprehension is possible because of relevant mental contents about the situation, the task, and the artefact. When people interact with an artefact, they create a dynamic representation that combines external information extracted through perceptual systems and knowledge in the human memory. This representation is used to simulate the artefact structure and functioning to enable reasoning processes (Cañas et al. 2001).

Comprehension is based on a good mental model of the technical artefact. However, it does not make sense to speak about the 'goodness of a mental model' before its information contents are known and its relevance concerning the artefact and task can be defined. Thinking is based on mental representations, and mental contents refer to their information contents. Thus, it is essential to discuss mental contents in mental models and their relation to human thinking.

The highest levels of cognitive processes are constructed by thinking and related processes. They also form the highest level of mental processes involved in HTI. Traditionally, thinking in psychology refers to such processes as categorization, inference, decision making, and problem solving (Newell and Simon 1972). In addition, it is essential to pay attention to representation construction, which underlies all types of thought processes (Saariluoma 2003). Much of the work in cognitive psychology that addresses lower cognitive processes has made a tacit assumption that cognitive capacity explains human performance (Broadbent 1958; Covan 2000; Miller 1956). However, the concept of cognitive capacity has limited explanatory power. When considering such issues as the correctness or relevance of information, content-based thinking can seldom be applied. The limited capacity of human information processing can entail any information. It can be correct or incorrect, as well as relevant or irrelevant. For this reason, the problems of relevance and correctness cannot be analysed in terms of capacity. Instead, new theoretical concepts are needed that are based on the notion of mental contents.

Mental contents—the content of the information in the human mind—are representations of the information that people use to guide their actions. They form their own type of scientific information (Saariluoma 2003) and open up theoretical and practical questions that are different from the perspectives discussed above. These questions are most relevant when higher cognitive processes such as apperception, consciousness, and thinking are considered (Newell and Simon 1972; Markman 1999; Saariluoma 1990, 1997, 2001). In HTI research, this means that mental contents must be examined when investigating interaction processes in which language, consciousness, and thinking play an essential role.

The world is depicted in people's minds. One can sit down on a chair if he or she has an image (i.e., a mental representation) of this chair in mind. This representation tells the individual's muscles when to contract and extend in order to maintain a correct sitting position. When sitting down, the individual is not very clearly aware of the strength of her muscles or the speed of the movement, but the information required for motion control (as well as all other activities) is stored in her memory. If a person does not have the information required for an action in his mind, he would not be able to perform the action in question. If the information were different in content, the action would be different as well.

The information in individuals' minds concerning the world, actions, goals, movements, beliefs, and other people is all entailed in human mental representations, which always have specific and definable information content that refers to something or tells about something. Psychological research that explains human behaviour based on the content of mental representations is called the *psychology of mental content*. This type of psychology strives to analyse the information content of human mental representations, and based on this information, to analyse and predict human behaviour (Saariluoma 1990, 2001, 2003).

Mental contents are important in the user psychology of thinking. Users piece together knowledge of their environment and the usage environment with the help of information contents in their memory (e.g., Nerb and Spada 2001). They control devices based on their own knowledge, which is represented as the content of their mental representations. If users lack the required information, they may use a device incorrectly or deficiently. For example, smartphone users may miss out on an essential service if they are unaware of it or do not know how to activate it (Kämäräinen and Saariluoma 2007). They lack the required mental contents to be able to use the given services.

Interaction designers should be aware of the beliefs, assumptions, wishes, ideas, needs, cultural engagements, and thoughts that underlie users' perceptions of the designed product. They should try to understand how users will most likely represent the product and the situation in question in their minds. Using question formulation of content-based psychology (Saariluoma 2003) and paying attention to the problems of first-time users can help identify this information.

To investigate the mind, it is necessary to elaborate on the idea of mental contents and to coin some additional theoretical concepts. The first thing to examine is perception. From a content-based point of view, perception cannot be understood as the sole process of constructing mental representations. Along with perception, many non-perceivable content elements also exist in mental representations. Leibniz (1704/1981) and Kant (1781/1976) were the first to notice this substantial conceptual problem over 200 years ago. Following their critical tradition, how mental representations are constructed with the help of non-perceivable content elements (i.e., *apperception*) will be examined next (Saariluoma 1990, 1992, 2003).

When using technical artefacts based on information technology, it is not always possible to see all the essential user interface components at a glance. Users have, for example, to navigate in menu systems, which hide the task goals under different menu bars, and imagine the hidden elements behind the actions of the menus. Drivers also have hidden elements when they operate a car. They have to be able to imagine—to represent in their mind—many issues that are not yet visible in the perceivable environment. For example, they seldom see their final destination at the start of the journey. They read traffic signs and follow the navigator's instructions and, on the basis of the given information, form mental representations of the destination.

Drivers use perceivable information elements, such as important buildings, traffic signs, and other landmarks, to form a representation of the situation. This representation of elements in time and space is the basis of *situation awareness* (Endsley 2006: Endsley and Jones 2011; Klein 2000), which is based on understanding the functional connection between the perceptual signs and the non-perceivable elements of mental representations (Saariluoma 1995). It thus includes the basic default that it is not possible to perceive all the contents of representational elements that effect thinking, and that not all fundamental elements that affect thinking are perceivable. Mental representations and analyses of situations have non-spatial and non-perceptual references, for example, such concepts as 'possible', 'friction', or 'eternity'.

The contents of mental representations are thus independent of the perceptual environment (Saariluoma 1990, 1995, 1997, 2001). It is possible to imagine electrons or think about foreign trade, for example, but it is not possible to directly perceive them. A person can perceive an individual computer or a human being but not the computer or the human in general. Accordingly, the supposition that representations are composed only through perception or attention should be abandoned. Instead, the concept of apperception can be used to explain the construction of representations in modern psychology (Saariluoma 1990, 1992, 1995, 2001, 2003; Saariluoma and Kalakoski 1997, 1998).

The basic mental function of apperception is constructing mental representations (Kant 1781/1976; Saariluoma 1990, 1992, 1995, 2001). Apperception selects and unifies information contents in the human mind as a coherent representation, thus forming interpretations of meaning for the physical world. It integrates conceptual information in the memory and sensory information, introduced by perceptual processes, into a representation that directs activity (Saariluoma 1990, 1995, 2001). Significant philosophers, such as Leibniz (1704/1981) and Kant (1781/1976), developed the concept of apperception. In psychology, particularly Wundt (1880) and Stout (1896) and later Saariluoma (1990, 1992, 1995, 2001) have used this concept. Consequently, an extensive amount of analytical work underlies the concept of apperception (Fig. 4.5).

Apperception unifies the memory representations with task- and situation-specific information contents. For example, medical doctors can understand the meaning of perceptual elements called symptoms and measurement values, as they can give these meaning in their mental representations. Yet non-medical observers can see perceptual elements but are unable to understand their medical significance. Thus, the same perceptual representations can have very different consequences; apperceptive processes give meaning to the perceptual elements.

Analysing mental contents provides an important perspective on human action and thinking (Saariluoma 2003), and is thus a key concept

Fig. 4.5 Apperception—two persons apperceive and mentally represent the same visual stimulus in a different way

in investigating HTI. It helps understand what kind of knowledge would be important and beneficial, and what kind of knowledge would be false and incorrect in designing easy-to-use user interfaces. Its task is also to expose implicit mental contents to explicit research.

To Decide and to Think

Psychologists consider and investigate thought processes that emerge in situations that have typical structures. For example, situations in which thinking produces two or more alternative courses of action are called *decision situations*, and the respective mental processes *decision processes* (Kahnemann 2011; Tversky and Kahnemann 1974, 1986). Likewise, *problem-solving situations* are those in which people have a goal, but do not know how to reach it by immediately available means (Newell and Simon 1972).

Decision making is straightforward in structure: one has to choose between two or more alternative courses of action. This activity is important in HTI, especially because machines cannot set goals. They are processes, by nature, with no independent goals. Artefacts are unable to know what is relevant. Designers and users are thus responsible for the human decision-making element: to organize the action of the technical artefacts and systems so that their behaviour makes sense from the human point of view (Kahnemann 2011).

In order to select goals and choose between different courses of usage action—that is to say, to make decisions—users have to collect and integrate information from different sources into a holistic representation of the situation. Decisions can be made alone or with other people, and they can be unified or conflict driven (Lehto and Nah 2006). In any case, the user has to make preferences among the possible alternatives.

On a formal level, decisions are supposed to be rational, which ideally means that the preferences attributed to the alternatives are perfect. In practice this is rare, as people tend to make the wrong decisions (Kahnemann 2011; Tversky and Kahnemann 1974). For a number of psychological reasons, they may overestimate the benefit they expect to receive from the chosen alternative. The challenge for designers is to ensure that people's mental representations are correct and that they have realistic expectations. This means that users' situation awareness should include adequate contents, which requires providing them with the necessary information, in the right form.

A tragic example of decision making is provided by an aeroplane accident in Madrid in 2008. The maintenance personnel of Spanair decided to disconnect a computer system that was about to crash. This system was supposed to be connected to a warning system that alerts the pilots of problems. During take-off the warning system was switched off, and the pilots were not aware that the flaps and slats were not well configured. As a consequence, they made a wrong decision when piloting the aircraft because they did not have all the relevant and necessary information.

Though decision making is an essential form of human thinking, when technical artefacts are used even more complex forms of thinking are needed. *Problem solving* is most important (Newell and Simon 1972). Complex problem solving is essential in designing technologies, but it also has a role in using technologies. In many professions, technology use includes complex tasks that require decision making, and carrying them out incorrectly may lead to serious problems.

Human beings are capable of constructing very large mental representations, for example, of cities, hospitals, banking systems, and aeroplanes. In all these cases, a large number of different content elements are combined to form a whole. Reaching goals in the numerous problem-solving and decision processes of the design of these entities presupposes good coordination. For example, a jumbo jet includes some 7 million different parts, and the human thought process enables its construction.

Human mental representations can be seen as large knowledge structures with contents that form coherent and consistent entities. In the same way that narrative analysis separates different parts of a story into independent wholes, the user interface designer must be able to distinguish between the essential content features of mental representations for scrutiny.

The analysis of human mental representations and their contents is not a mechanical process. Systems of mental contents are largely unknown even to the object of observation. They contain a multitude of subconscious layers, which usability research seeks to understand. For instance, response to interview questions cannot be taken at face value; one also has to be able to read the features of representations that crop up between the lines. Thus, basing usability analysis on personal experiences is misguiding. Examining these may be inspiring, but they prove rather little about the issue as a whole. The subconscious can normally be reached through the point of view of an external 'third-person' perspective.

In order to understand representation, its individual content elements should be detached from the whole and described, just as micro-organisms separated from other microscopic mass are pictured in bacteria research. The limits of objects deemed essential should be accurately defined and delineated; afterwards, it is possible to draw conclusions about their meaning. One perspective on how the contents of mental representations control humans when interacting with technology can be seen by looking at the similarities and dissimilarities in the contents of representations.

The ability to solve problems is also needed in unpredictable contextual conditions. Technologies can behave in ways that are impossible to predict, and users may be forced to solve important problems in order to make the right decisions when using technologies. The Fukushima nuclear power plant was built by the sea in order to access the large amount of water needed to cool the reactors. However, the designers did not envision the possibility that a tsunami and an earthquake would hit the nuclear power plant. This is an example of how designers should 'design for resilience': to be prepared for the unexpected. Understanding the nature of human problem solving is the basis for designing any HTI. Designing for resilience makes this demand a prerequisite.

Cognitive Processes and Effortless Interaction

In conclusion, this chapter has focused on the information required to make it possible and easy for people to use technical artefacts. The core of the discussion has focused on the psychology behind fluent use. The central message is that merely generating event sequences and other technical solutions is insufficient to design successful user interfaces; the preconditions for smooth and unproblematic interactions must be understood. This chapter has thus focused on analysing basic cognitive processes when interacting with technologies. From the point of view of interaction design, emotions, individual differences, and groups and cultures also play an important role in successful HTI design. These will be discussed in the forthcoming chapters.

It is essential that people are able to use new technologies. Thus, a well-designed artefact must work appropriately *and* be highly usable. As has been illustrated here, human psychological (and in particular, cognitive) processes set different types of limits that have to be taken into account when designing effective user interfaces.

Designing usable systems presupposes solving several levels of problems that arise from the nature of the human mind. Artefacts do not cause them, because there are many ways of developing user interfaces between man and machines, and it is human beings who create artefacts. Although it is not always easy to understand the human mind and how it functions, it should be possible to create successful user interfaces by taking into account the characteristics of human perception, memory, decision-making, and thinking processes.

This chapter has considered the necessity of being in perceptual and psychomotor contact with artefacts, entailing such issues as responding and controlling. Information about a *perceptual-motor cycle* provides tools for examining these basic issues. The necessity of *learning and remembering* raises another set of important cognitive issues concerning immediate interaction. People have to be able to learn and remember the necessary information in order to use technical artefacts. Designing the best conditions for these processes requires applying the psychology of learning and memory.

The next set of cognitive questions is intimately connected to human memory. It concerns *communication and comprehension*. Designers create sets of signs to be able to design sufficiently extensive variations when using technologies. For example, programming languages provide the means to significantly vary the behaviour of technical artefacts. Finally, various types of thought processes, such as decision making and complex problem solving, are relevant in HTI. It is often too easily seen that technologies are no more than simple stimulus-process artefacts that do not require creative thinking to use. But even such a basic technology as a chisel, for example, can be used to complete such complex tasks as creating a sculpture. The psychology of 'being able to use' introduces a new set of theoretical concepts to interaction analysis, which are mostly related to psychology, but in some cases the study of semiotics and the philosophy of the mind can be very beneficial.

References

Anderson, J. R. (1976). *Language, memory and thought*. Hillsdale, NJ: Erlbaum.

Anderson, J. R. (1983). *The architecture of cognition*. Cambridge, MA: Harvard University Press.

Anderson, J. R. (1993). *Rules of the mind*. Hillsdale, NJ: Erlbaum.

Anderson, J. R., Farrell, R., & Sauers, R. (1984). Learning to program Lisp. *Cognitive Science, 8*, 87–129.

Atkinson, R., & Shiffrin, R. (1968). Human memory: A proposed system. In K. W. Spence & J. T. Spence (Eds.), *The psychology of learning and motivation* (Vol. 2, pp. 89–195). New York: Academic Press.

Baccino, T., & Manunta, Y. (2005). Eye-fixation-related potentials: Insight into parafoveal processing. *Journal of Psychophysiology, 19*, 204–215.

Baddeley, A. D. (1986). *Working memory*. Cambridge: Cambridge University Press.

Baddeley, A. (1990). *Human memory: Theory and practice*. Hillsdale, NJ: Erlbaum.

Baddeley, A. (2007). *Working memory, thought, and action*. Oxford, UK: Oxford University Press.

Baddeley, A. (2012). Working memory: Theories, models, and controversies. *Annual Review of Psychology, 63*, 1–29.

Barnaby, J. (1980). Lighting for productivity gains. *Lighting Design Application, 10*, 20–28.

Barnard, P., & Grudin, J. (1988). Command names. In M. Helander (Ed.), *Handbook of human-computer interaction* (pp. 237–255). Amsterdam: North-Holland.

Beck, U. (1992). *Risk society: Towards a new modernity*. London: Sage.

Beck, U. (2008). *Weltrisikogesellschaft* [World risk society]. Frankfurth am Main: Surkamp.

Blackmon, M. H., Kitajima, M., & Polson, P. G. (2005). Tool for accurately predicting website navigation problems, non-problems: Problem severity, and effectiveness of repairs. In *CHI'05 Proceedings of the SIGCHI Conference on Human Factors in Computing Systems* (pp. 31–40).

Bonnefond, A., Muzet, A., Winter-Dill, A., Bailloeuil, C., Bitouze, F., & Bonneau, A. (2001). Innovative working schedule: Introducing one short nap during the night shift. *Ergonomics, 44*, 937–945.

Bouma, H., Fozard, J. L., & van Bronswijk, J. E. M. H. (2009). Gerontechnology as a field of endeavour. *Gerontechnology, 8*, 68–75.

Bower, G. H. (1970). Organizational factors in memory. *Cognitive Psychology, 1*, 18–46.

Bower, G. H. (1981). Mood and memory. *American Psychologist, 36*, 129–148.

Bower, G. H., Clark, M. C., Lesgold, A. M., & Winzenz, D. (1969). Hierarchical retrieval schemes in recall of categorized word lists. *Journal of Verbal Learning and Verbal Behavior, 8*, 323–343.

Boyton, R. M. (1988). Color vision. *Annual Review of Psychology, 39*, 69–100.

Brady, T. F., Konkle, T., Oliva, A., & Alvarez, G. A. (2009). Detecting changes in real-world objects: The relationship between visual long-term memory and change blindness. *Communicative and Integrative Biology, 2*, 1–3.

Bregman, A. S. (1994). *Auditory scene analysis: The perceptual organization of sound*. Cambridge, MA: MIT Press.

Bridger, R. S. (2009). *Introduction to ergonomics*. Boca Raton, FL: CRC Press.

Broadbent, D. (1958). *Perception and communication.* London: Pergamon Press.

Broadbent, D. (1975). The magic number seven after twenty years. In R. Kennedy & A. Wilkes (Eds.), *Studies in long term memory* (pp. 253–287). New York: Wiley.

Brooks, R. (1980). Studying programmer behaviour experimentally: The problem of proper methodology. *Communications of the ACM, 23,* 207–213.

Brown, R. (2000). *Group processes.* Oxford: Basil Blackwell.

Brown, J. R., Earnshaw, R., Jern, M., & Vince, J. (1995). *Visualization: Using computer graphics to explore data and present information.* Hoboken, NJ: Wiley.

Bruce, V., & Green, P. R. (1992). *Visual perception: Physiology, psychology and ecology.* Hove: Psychology Press.

Bruce, N. D., & Tsotsos, J. K. (2009). Saliency, attention, and visual search: An information theoretic approach. *Journal of Vision, 9,* 1–24.

Cañas, J. J., Antolí, A., & Quesada, J. F. (2001). The role of working memory on measuring mental models of physical systems. *Psicológica, 22,* 25–42.

Card, S., Moran, T., & Newell, A. (1983). *The psychology of human-computer interaction.* Hillsdale, NJ: Erlbaum.

Chandler, D. (2007). *Semiotics: The basics.* London: Routledge.

Charness, N. (1976). Memory for chess positions: Resistance to interference. *Journal of Experimental Psychology: Human Learning and Memory, 2,* 641–653.

Charness, N. (2009). Ergonomics and aging: The role of interactions. In I. Graafmans, V. Taipale, & N. Charness (Eds.), *Gerontechnology: Sustainable investment in future* (pp. 62–73). Amsterdam: IOS Press.

Chase, W. G., & Ericsson, K. A. (1981). Skilled memory. In J. Andersson (Ed.), *Cognitive skills and their acquisition* (pp. 141–189). Hillsdale, NJ: Erlbaum.

Chun, M. M., Golomb, J. D., & Turk-Browne, N. B. (2011). A taxonomy of external and internal attention. *Annual Review of Psychology, 62,* 73–101.

Cockton, G. (2004). Value-centred HCI. In *Proceedings of the Third Nordic Conference on Human-Computer Interaction* (pp. 149–160).

Covan, N. (1999). An embedded-process model of working memory. In A. Miyake & P. Shah (Eds.), *Models of working memory.* Cambridge: Cambridge University Press.

Covan, N. (2000). The magical number 4 in short-term memory: A reconsideration of mental storage capacity. *Behavioural and Brain Sciences, 24,* 87–185.

Cronbach, L. J. (1984). *Essentials of psychological testing.* New York: Harper-Collins.

Crowder, R. (1976). *The structure of human memory*. Hillsdale, NJ: Erlbaum.

Czaja, S. J., & Nair, S. N. (2006). Human factors engineering and systems design. In G. Salvendy (Ed.), *Handbook of human factors and ergonomics* (pp. 32–49). Hoboken, NJ: Wiley.

de Souza, C. S. (2005). *The semiotic engineering of human-computer interaction*. Cambridge, MA: MIT Press.

Desmet, P., Overbeeke, K., & Tax, S. (2001). Designing products with added emotional value: Development and application of an approach for research through design. *The Design Journal, 4*, 32–47.

Desmurget, M., Epstein, C., Turner, R., Prablanc, C., Alexander, G. E., & Grafton, S. T. (1999). Role of the posterior parietal cortex in updating reaching movements to a visual target. *Nature Neuroscience, 2*, 563–567.

Dillon, A., Richardson, J., & McKnight, C. (1990). The effects of display size and text splitting on reading lengthy text from screen. *Behaviour and Information Technology, 9*, 215–227.

Duncan, J., & Humphreys, G. W. (1989). Visual search and stimulus similarity. *Psychological Review, 96*, 433–458.

Eco, U. (1976). *A theory of semiotics*. Bloomington, IN: Indiana University Press.

Egeth, H. E., & Yantis, S. (1997). Visual attention: Control, representation, and time course. *Annual Review of Psychology, 48*, 269–297.

Eimer, M., Nattkemper, D., Schröger, E., & Printz, W. (1996). Involuntary attention. In O. Neuman & A. F. Sanders (Eds.), *Handbook of perception and action 3. Attention* (pp. 155–184). London: Academic Press.

Endsley, M. R. (1995). Toward a theory of situation awareness in dynamic systems. *Human Factors: The Journal of the Human Factors and Ergonomics Society, 37*, 32–64.

Endsley, M. R. (2006). Expertise and situation awareness. In K. A. Ericsson, N. Charness, P. J. Feltovich, & R. Hoffman (Eds.), *The Cambridge handbook of expertise and expert performance* (pp. 633–651). Cambridge: Cambridge University Press.

Endsley, M., & Jones, D. (2011). *Designing situation awareness*. Boca Raton, FL: CRC Press.

Ericsson, K. A. (2006). The influence of experience and deliberate practice on the development of superior expert performance. In K. A. Ericsson, N. Charness, P. J. Feltovich, & R. Hoffman (Eds.), *The Cambridge handbook of expertise and expert performance* (pp. 683–704). Cambridge: Cambridge University Press.

Ericsson, K. A., & Kintsch, W. (1995). Long-term working memory. *Psychological Review, 102*, 211–245.

Forrester, M., & Reason, D. (1990). HCI 'Intraface Model' for system design. *Interacting with Computers, 2*, 279–296.

Fotowat, H., & Gabbiani, F. (2011). Collision detection as a model for sensory-motor integration. *Annual Review of Neuroscience, 34*, 1–19.

Funke, J. (2012). Complex problem solving. In N. M. Seel (Ed.), *Encyclopedia of the sciences of learning* (pp. 682–685). Heidelberg: Springer.

Gallese, V., & Lakoff, G. (2005). The brain's concepts: The role of the sensory-motor system in conceptual knowledge. *Cognitive Neuropsychology, 22*, 455–479.

Gentner, D., & Stevens, A. L. (1983). *Mental models.* Hillsdale, NJ: Erlbaum.

Gibson, J. J. (1950). *The perception of the visual world.* Boston, MA: Houghton Mifflin.

Gibson, J. J. (1979). *The ecological approach to visual perception.* Boston, MA: Houghton Mifflin.

Gielo-Perczak, K., & Karwowski, W. (2003). Ecological models of human performance based on affordance, emotion and intuition. *Ergonomics, 46*, 310–326.

Gigerenzer, G., & Gaissmaier, W. (2011). Heuristic decision making. *Annual Review of Psychology, 62*, 451–482.

Gobet, F. (2000). Some shortcomings of long-term working memory. *British Journal of Psychology, 91*(4), 551–570.

Goldstein, E. B. (2010). *Sensation and perception.* Belmont, CA: Wadsworth.

Gopher, D., & Donchin, E. (1986). Workload: An examination of the concept. In K. R. Boff, L. Kaufman, & J. P. Thomas (Eds.), *Handbook of perception and human performance: Cognitive processes and performance* (pp. 1–46). Hoboken, NJ: Wiley.

Green, T. R. G., & Petre, M. (1996). Usability analysis of visual programming environments: A 'cognitive dimensions' framework. *Journal of Visual Languages and Computing, 7*, 131–174.

Gregg, V. (1986). *Introduction to human memory.* London: Routledge.

Häkkinen, M. (2010). Why alarms fail: A cognitive explanatory model. In *Jyväskylä Studies in Computing* (Vol. 127). Jyväskylä: Jyväskylä University Press.

Hassenzahl, M. (2011). *Experience design.* San Rafael, CA: Morgan & Claypool.

Hecht, S. (1924). The visual discrimination of intensity and the Weber-Fechner law. *The Journal of General Physiology, 7*, 235–267.

Helfenstein, S., & Saariluoma, P. (2006). Mental contents in transfer. *Psychological Research, 70*, 293–303.

Henderson, J. M., & Hollingworth, A. (1999). High-level scene perception. *Annual Review of Psychology, 50*, 245–271.

Howarth, P. A. (1999). Oculomotor changes within virtual environments. *Applied Ergonomics, 30*, 59–67.

Isler, R., Kirk, P., Bradford, S. J., & Parker, R. J. (1997). Testing the relative conspicuity of safety garments for New Zealand forestry workers. *Applied Ergonomics, 28*, 323–329.

James, W. (1890). *The principles of psychology*. New York: Dover.

Jeannerod, M. (2007). *Motor cognition*. Oxford: Oxford University Press.

Johnson-Laird, P. (1983). *Mental models: Towards a cognitive science of language, inference, and consciousness*. Cambridge, MA: Harvard University Press.

Johnson-Laird, P. (2008). *How we reason?* Oxford: Oxford University Press.

Josephson, S., & Holmes, M. E. (2002). Visual attention to repeated internet images: Testing the scanpath theory on the world wide web. In *Proceedings of the 2002 Symposium on Eye Tracking Research & Applications* (pp. 43–49). New York: ACM Press.

Judd, C. H. (1902). Practice and its effects on the perception of illusions. *Psychological Review, 9*, 27–39.

Kahnemann, D. (1973). *Attention and effort*. Englewood Cliffs, NJ: Prentice-Hall.

Kahnemann, D. (2011). *Thinking, fast and slow*. London: Penguin Books.

Kalakoski, V., & Saariluoma, P. (2001). Taxi drivers' exceptional memory of street names. *Memory and Cognition, 29*, 634–638.

Kämäräinen, A., & Saariluoma, P. (2007). Under-use of mobile services: How advertising space is used. In *Proceedings of the 9th International Workshop on Internationalization of Products and Systems* (pp. 19–29).

Kant, I. (1781/1976). *Kritik der reinen Vernunft* [The critique of pure reason]. Hamburg: Felix Meiner.

Karwowski, W. (2006). The discipline of ergonomics and human factors. In G. Salvendy (Ed.), *Handbook of human factors and ergonomics* (pp. 3–31). Hoboken, NJ: Wiley.

Kawato, M., Furukawa, K., & Suzuki, R. (1987). A hierarchical neural-network model for control and learning of voluntary movement. *Biological Cybernetics, 57*, 169–185.

Kintsch, W., & Van Dijk, T. A. (1978). Toward a model of text comprehension and production. *Psychological Review, 85*, 363–394.

Kitajima, M., & Polson, P. G. (1995). A comprehension-based model of correct performance and errors in skilled, display-based, human-computer interaction. *International Journal of Human-Computer Studies, 43*, 65–99.

Kivimäki, M., & Lindström, K. (2006). Psychosocial approach to occupational health. In G. Salvendy (Ed.), *Handbook of human factors and ergonomics* (pp. 3–31). Hoboken, NJ: Wiley.

Klein, G. (2000). Analysis of situation awareness from critical incident reports. In M. R. Endsley & D. J. Garland (Eds.), *Situation awareness analysis and measurement* (pp. 51–71). Boca Raton, FL: CRC Press.

Kline, P. (2013). *The handbook of psychological testing*. London: Routledge.

Koelega, H. S. (1996). Sustained attention. In O. Neuman & A. F. Sanders (Eds.), *Handbook of perception and action 3: Attention* (pp. 277–331). London: Academic Press.

Lamble, D., Kauranen, T., Laakso, M., & Summala, H. (1999). Cognitive load and detection thresholds in car following situations: Safety implications for using mobile (cellular) telephones while driving. *Accident Analysis and Prevention, 31*, 617–623.

Land, M. F., & McLeod, P. (2000). From eye movements to actions: How batsmen hit the ball. *Nature Neuroscience, 3*, 1340–1345.

Landauer, T. (2007). LSA as a theory of meaning. In T. Landauer, D. McNamara, S. Dennis, & W. Kintsch (Eds.), *Handbook of latent semantic analysis* (pp. 3–34). Mahwah, NJ: Erlbaum.

Landauer, T. K., & Dumais, S. T. (1997). A solution to Plato's problem: The latent semantic analysis theory of acquisition, induction, and representation of knowledge. *Psychological Review, 104*, 211–240.

Larkin, J. H., McDermott, J., Simon, D., & Simon, H. A. (1980). Expert and novice performance in solving physics problems. *Science, 208*, 1335–1342.

Laursen, L. H., Hansen, H. L., & Jensen, O. C. (2008). Fatal occupational accidents in Danish fishing vessels 1989–2005. *International Journal of Injury Control and Safety Promotion, 15*(2), 109–117.

Lehto, M. R., & Nah, F. (2006). Decision making models and decision support. In G. Salvendy (Ed.), *Handbook of human factors and ergonomics* (pp. 191–242). Hoboken, NJ: Wiley.

Leibniz, G. (1704/1981). *New essays on human understanding*. Cambridge: Cambridge University Press.

Leikas, J. (2009). *Life-based design—A holistic approach to designing human-technology interaction*. Helsinki: Edita Prima Oy.

Lewis, J. R., Potosnak, K. M., & Magyar, R. L. (1997). Keys and keyboards. In H. Helander, T. Landauer, & P. Prabhu (Eds.), *Handbook of human-computer interaction* (pp. 1285–1315). Amersterdam: Elsevier.

Liu, H., Selker, T., & Lieberman, H. (2003). Visualizing the affective structure of a text document. CHI'03 Extended Abstracts on Human Factors in Computing Systems, 740-741.

Logie, R. H. (1995). *Visuo-spatial working memory*. Hove: Psychology Press.

Luckiesh, M. (1965). *Visual illusions*. New York: Dover.

Lyons, J. (1977). *Semantics* (Vols. 1–2). Cambridge: Cambridge University Press.

Mackworth, N. H., & Morandi, A. J. (1967). The gaze selects informative details within pictures. *Perception and Psychophysics, 2*, 547–552.

Markman, A. (1999). *Knowledge representation*. Mahwah, NJ: Lawrence Erlbaum.

Mayer, R. (1997). From novice to expert. In H. Helander, T. Landauer, & P. Prabhu (Eds.), *Handbook of human-computer interaction* (pp. 781–797). Amsterdam: North-Holland.

McKeithen, K., Reitman, J., Rueter, H., & Hirtle, S. (1981). Knowledge organization and skill differences in computer programmers. *Cognitive Psychology, 13*, 307–325.

Miller, G. A. (1956). The magical number seven, plus or minus two: Some limits on our capacity for processing information. *Psychological Review, 63*, 81–97.

Minsky, M. L. (1967). *Computation: Finite and infinite machines*. Englewood Cliffs, NJ: Prentice-Hall.

Mollon, J. D. (1982). Color vision. *Annual Review of Psychology, 33*, 41–85.

Moran, T. P. (1981). Guest editor's introduction: An applied psychology of the user. *ACM Computing Surveys, 13*, 1–11.

Murata, A., Uetake, A., Matsumoto, S., & Takasawa, Y. (2003). Evaluation of shoulder muscular fatigue induced during VDT tasks. *International Journal of Human-Computer Interaction, 15*, 407–417.

Nasar, J., Hecht, P., & Wener, R. (2008). Mobile telephones, distracted attention, and pedestrian safety. *Accident Analysis and Prevention, 40*, 69–75.

Navarro-Prieto, R., & Canas, J. (2001). Are visual programming languages better? The role of imagery in program comprehension. *International Journal of Human-Computer Studies, 54*, 799–829.

Neisser, U. (1963). Decision time without reaction time. *American Journal of Psychology, 76*, 376–385.

Neisser, U. (1967). *Cognitive psychology*. New York: Appleton-Century-Crofts.

Neisser, U. (1976). *Cognition and reality*. San Francisco, CA: Freeman.

Nerb, J., & Spada, H. (2001). Evaluation of environmental problems: A coherence model of cognition and emotion. *Cognition and Emotion, 15*, 521–551.

Neumann, O. (1984). Automatic processing: A review of recent findings and a plea for an old theory. In W. Prinz & A. Sanders (Eds.), *Cognition and motor processes* (pp. 255–293). Berlin: Springer.

Newell, A., & Rosebloom, P. (1981). Mechanisms of skills acquisition and the law of practice. In J. R. Anderson (Ed.), *Cognitive skills and their acquisition* (pp. 1–55). Hillsdale, NJ: Erlbaum.

Newell, A., & Simon, H. A. (1972). *Human problem solving*. Engelwood Cliffs, NJ: Prentice-Hall.

Nielsen, J. (1993). *Usability engineering*. San Diego, CA: Academic Press.

Norman, D. A. (1999). Affordance, conventions, and design. *Interactions, 6*, 38–43.

Norman, D. A. (2002). *The design of everyday things*. New York: Basic Books.

Norman, D. (2004). *Emotional design: Why we love (or hate) everyday things*. New York: Basic Books.

Ogden, C. & Richards, I. (1923). *The Meaning of Meaning*. London: Routledge & Kegan Paul.

Olson, G. M., & Olson, J. S. (2003). Human-computer interaction: Psychological aspects of human use of computing. *Annual Review of Psychology, 54*, 491–516.

Oulasvirta, A., & Saariluoma, P. (2004). Long-term working memory and interrupting messages in human–computer interaction. *Behaviour and Information Technology, 23*, 53–64.

Oulasvirta, A., & Saariluoma, P. (2006). Surviving task interruptions: Investigating the implications of long-term working memory theory. *International Journal of Human-Computer Studies, 64*, 941–996.

Parasuraman, R., & Rizzo, M. (2006). *Neuroergonomics: The brain at work*. Oxford: Oxford University Press.

Pashler, H. (1998). *The psychology of attention*. Cambridge, MA: MIT Press.

Pashler, H., Johnston, J. C., & Ruthruff, E. (2001). Attention and performance. *Annual Review of Psychology, 52*, 629–651.

Pavlov, I. (1927). *Conditioned reflexes*. New York: International Publishers.

Pennington, N. (1987). Stimulus structures and mental representations in expert comprehension of computer programs. *Cognitive Psychology, 19*, 295–341.

Pinch, T. J., & Bijker, W. E. (1984). The social construction of facts and artefacts: Or how the sociology of science and the sociology of technology might benefit each other. *Social Studies of Science, 14*, 399–441.

Posner, M. I. (1980). Orienting of attention. *Quarterly Journal of Experimental Psychology, 32*, 3–25.

Proctor, R. W., & Proctor, J. D. (2006). Sensation and perception. In G. Salvendy (Ed.), *Handbook of human factors and ergonomics* (pp. 53–88). Hoboken, NJ: Wiley.

Ramsay, M., & Nielsen, J. (2000). WAP usability, Déjà Vu: 1994 all over again. *Report from a Field Study in London.* Nielsen Norman Group.

Rasmussen, J. (1983). Skills, rules, and knowledge; signals, signs, and symbols, and other distinctions in human performance models. *Systems, Man and Cybernetics, IEEE Transactions, 13*(3), 257–266.

Rasmussen, J., & Rouse, W. B. (1981). *Human detection and diagnosis of system failures.* New York: Plenum.

Rayner, K. (1998). Eye movements in reading and information processing: 20 years of research. *Psychological Bulletin, 124*, 372–422.

Rieman, J., Young, R. M., & Howes, A. (1996). A dual-space model of iteratively deepening exploratory learning. *International Journal of Human-Computer Studies, 44*, 743–775.

Robins, A., Rountree, J., & Rountree, N. (2003). Learning and teaching programming: A review and discussion. *Computer Science Education, 13*, 137–172.

Rock, I. (1983). *The logic of perception.* Cambridge, MA: MIT Press.

Rosson, B., & Carroll, J. (2002). *Usability engineering: Scenario-based development of human-computer interaction.* San Francisco, CA: Morgan Kaufmann.

Rumelhart, D. E., & Norman, D. A. (1982). Simulating a skilled typist: A study of skilled cognitive-motor performance. *Cognitive Science, 6*, 1–36.

Saariluoma, P. (1990). Apperception and restructuring in chess players' problem solving. In K. J. Gilhooly, M. T. G. Keane, R. H. Logie, & G. Erdos (Eds.), *Lines of thought: Reflections on the psychology of thinking* (pp. 41–57). London: Wiley.

Saariluoma, P. (1992). Error in chess: Apperception restructuring view. *Psychological Research, 54*, 17–26.

Saariluoma, P. (1995). *Chess players' thinking.* London: Routledge.

Saariluoma, P. (1997). *Foundational analysis: Presuppositions in experimental psychology*. London: Routledge.

Saariluoma, P. (2001). Chess and content-oriented psychology of thinking. *Psicologica, 22*, 143–164.

Saariluoma, P. (2003). Apperception, content-based psychology and design. In U. Lindeman (Ed.), *Human behavior in design* (pp. 72–78). Berlin: Springer.

Saariluoma, P. (2005). Explanatory frameworks for interaction design. In A. Pirhonen, H. Isomäki, C. Roast, & P. Saariluoma (Eds.), *Future interaction design* (pp. 67–83). London: Springer.

Saariluoma, P., & Kalakoski, V. (1997). Skilled imagery and long-term working memory. *American Journal of Psychology, 110*, 177–201.

Saariluoma, P., & Kalakoski, V. (1998). Apperception and imagery in blindfold chess. *Memory, 6*, 67–90.

Saariluoma, P., & Oulasvirta, A. (2010). User psychology: Re-assessing the boundaries of a discipline. *Psychology, 1*, 317–328.

Saariluoma, P., & Rousi, R. (2015). Symbolic interactions: Towards a cognitive scientific theory of meaning in human technology. *Journal of Advances in Humanities, 3*, 310–323.

Saariluoma, P., & Sajaniemi, J. (1989). Visual information chunking in spreadsheet calculation. *International Journal of Man–Machine Studies, 30*, 475–488.

Saariluoma, P., & Sajaniemi, J. (1994). Transforming verbal descriptions into mathematical formulas in spreadsheet calculation. *International Journal of Human-Computer Studies, 41*, 915–948.

Salmeron, L., Canas, J., Kintsch, W., & Fajardo, I. (2005). Reading strategies and hypertext comprehension. *Discourse Processes, 40*, 171–191.

Salthouse, T. A. (1986). Perceptual, cognitive, and motoric aspects of transcription typing. *Psychological Bulletin, 99*, 303–319.

Sanders, A. F., & Donk, M. (1996). Visual search. In O. Neuman & A. Sanders (Eds.), *Handbook of perception and action* (Vol. II, pp. 43–77). London: Academic Press.

Sassen, S. (2002). Towards a sociology of information technology. *Current Sociology, 50*, 365–388.

Schmidt, R. A. (1975). A schema theory of discrete motor skill learning. *Psychological Review, 82*, 225–260.

Schmidt, R.A., & Lee, T.D. (2011). *Motor control and learning: A behavioral emphasis*. (5th ed). Champaign, IL: Human Kinetics.

Schneider, W., Dumais, S., & Shiffrin, R. (1984). Automatic and controlled processing and attention. In R. Parasuraman & D. Davies (Eds.), *Varieties of attention*. Orlando, FL: Academic Press.

Searle, J. (1993). *Intentionality*. Cambridge: Cambridge University Press.

Shepard, R. N. (1967). Recognition memory for words, sentences, and pictures. *Journal of Verbal Learning and Verbal Behavior, 6*, 156–163.

Shneiderman, B. (2011). Tragic errors: Usability and electronic health records. *Interactions, 18*, 60–63.

Shneiderman, B., & Plaisant, C. (2005). *Designing user interfaces*. Boston, MA: Pearson.

Simon, H. A. (1955). A behavioural model of rational choice. In H. Simon (Ed.), *Models of thought* (pp. 7–19). New Haven, CT: Yale University Press.

Singley, M. K., & Anderson, J. R. (1987). A keystroke analysis of learning and transfer in text editing. *Human-Computer Interaction, 3*, 223–274.

Soto, R. (1999). Learning and performing by exploration: Label quality measured by latent semantic analysis. In *Proceedings of the SIGCHI Conference on Human Factors in Computing Systems* (pp. 418–425).

Standing, L. (1973). Learning 10,000 pictures. *Quarterly Journal of Experimental Psychology, 25*, 207–222.

Standing, L., Conezio, J., & Harber, R. N. (1970). Perception and memory for pictures: Single trial learning of 2560 stimuli. *Psychonomic Science, 19*, 73–74.

Stout, G. (1896). *Analytic psychology*. London: Routledge.

Styles, E. (1997). *The psychology of attention*. Hove: Psychology Press.

Thorndike, E., & Woodworth, R. (1901). The influence of improvement in one mental function upon the efficiency of other functions. *Psychological Review, 8*, 247–261.

Treisman, A., & Gelade, G. (1980). A feature integration theory of attention. *Cognitive Psychology, 12*, 97–136.

Tufte, E. R. (1990). *Envisioning information*. Cheshire: Graphics Press.

Tufte, E. R. (1997). *Visual explanation*. Cheshire: Graphics Press.

Turing, A. M. (1936–1937). On computable numbers, with an application to the entscheidungsproblem. *Proceedings of the London Mathematical Society, 42*, 230–265.

Tversky, A., & Kahneman, D. (1986). Rational choice and the framing of decisions. *Journal of Business, 59*, 251–278.

Tversky, A., & Kahnemann, D. (1974). Judgement under uncertainty: Heuristics and biases. Science, 185, 1124–31.

Ullman, S. (1996). *High level vision*. Cambridge, MA: MIT Press.

van der Heijden, A. H. C. (1992). *Selective attention in vision*. London: Routledge.

van der Heijden, A. H. C. (1996). Visual attention. In O. Neuman & A. F. Sanders (Eds.), *Handbook of perception and action 3. Attention*. London: Academic Press.

Visser, W., & Hoc, J. M. (1990). Expert software design strategies. In J. M. Hoc, T. R. G. Green, R. Samurcay, & D. Gilmore (Eds.), *Psychology of programming* (pp. 235–249). New York: Academic Press.

Weiser, M. (1993). Some computer science issues in ubiquitous computing. *Communications of the ACM, 36*, 75–84.

Wickens, C., & Holands, J. G. (2000). *Engineering psychology and human performance*. Upper Saddle River, NJ: Prentice-Hall.

Wundt, W. (1880). *Grundriss der Psychologie* [Foundations of psychology]. Stuttgart: Kröner.

Yarbus, A. L. (1967). *Eye movements and vision*. New York: Plenum Press.

5

Emotions, Motives, Individuals, and Cultures in Interaction

Cognitive aspects of the human mind form the foundations of solving usability problems. However, being able to use a technology is not the only critical psychological question in the design of successful HTI. In addition to understanding users' capabilities, it is equally important to comprehend their preferences and what they want to accomplish with the help of technologies. Knowledge of the dynamic mind—in particular human emotions, motives, and personality—helps address such 'liking and wanting' concerns in HTI design.

Though people can use a given technology, such as an app in a smartphone or a medical instrument, it does not yet mean they *like* to use it. It may be that they simply discard it as quickly as possible and replace it with another because they are not motivated to use it. It may also be that the designed technology does not attract the target audience, even though objectively they need it. People may find the technology ugly, unfashionable or even stigmatizing, and therefore do not *want* to adopt it at all. It may also be that the technology does not fit in with peoples' values, self-image or lifestyle, and is therefore avoided. These questions of *like* and *want*, which concern personal and group emotions and motives, introduce a number of new design discourses (Brave and Nass 2009;

© The Editor(s) (if applicable) and The Author(s) 2016
P. Saariluoma et al., *Designing for Life*,
DOI 10.1057/978-1-137-53047-9_5

Hassenzahl 2011; Helander and Khalid 2006; Nagamashi 2011; Norman 2004). These discourses as a whole can be called *dynamic user psychology* (*emotions, motives, and personality*), as their scientific background is in dynamic psychology.

The influence of emotions, motives, and personality should not be underestimated, as it can affect the choices of a large group of people, and even of society. For example, after the Three Mile Island (1979), Chernobyl (1986), Fosmark (2006), and Fukushima (2011) accidents, people felt unsafe and were uncertain about nuclear energy, and partly lost their trust in this type of technology (Fig. 5.1). People's situational assessments changed as the risk became visible.

Fig. 5.1 After nuclear accidents people lost their trust in this type of technology

The headlines in the media, and concrete actions such as limiting the distribution of milk or the recommendation to avoid eating fish, made the consequences of the accidents concrete in everyday life. This concreteness is essential in changing situational assessments and attitudes (Kahnemann 2011; Tversky and Kahnemann 1974). In the case of Fukushima, the concreteness of the effects was further intensified by social media (Friedman 2011). People no longer found nuclear energy a good solution and fought against it. This resistance, caused by changing opinions, finally forced industry and many governments to take a more critical position towards nuclear energy.

Questions of likes and wants are vital in understanding why people accept and adopt new technologies. They are intimately related to emotions, motives, and personality, which are the basic psychological concepts that explain the issues of acceptance and adoption (Brave and Nass 2009; Hassenzahl 2011; Nagamashi 2011; Norman 2004). These aspects of the human mind are considered under the concepts of dynamic psychology to separate them from cognitive issues (Neisser 1967).

Historically, the concept of dynamic psychology or psychodynamics arises from the clinical personality theories of Freud (1917/2000) and Jung (1999). Since both of these luminaries are psychoanalysts, it has occasionally been said that the term 'psychodynamic' refers to psychoanalysis. This usage of the word 'dynamic' in psychology nevertheless misses the fact that many other psychologists have also worked to improve understanding of emotions, motives, and personality (Neisser 1967). The most recent paradigm of dynamic psychology is *positive psychology*. It focuses on such issues as well-being, contentment, satisfaction, happiness, and the flow of experience, which have proven to be important in many contexts of *user experience* (Seligman and Csikszentmihalyi 2000).

This chapter uses dynamic psychology to include the psychology of emotions, motives, personality, and intercultural issues. The concept of dynamic refers here to the 'forces' that control and direct human actions, but which also give them 'mental energy'. These make people act in a defined way and give them the power to meet and overcome different obstacles.

The main issues discussed in this chapter are:

- User experience and emotional usability;
- Motivation and the will to use;
- Individuals and personality; and
- Interacting groups and cultures.

From a metascientific point of view, the task of interaction research is to find rational grounds for solving interaction design problems. This presupposes investigation of general psychological phenomena and connecting this knowledge to finding and grounding solutions for technological problems. Following this line of thinking, it is possible to discuss on a scientific level the role of the psychology of dynamic interaction (i.e., the roles of emotions, motives, and personality in HTI).

Emotions and the Mind

Emotions form a central dimension of the human mind (Damasio 2005; LeDoux 1998; Rolls 2000). A better understanding of emotional processes in the analyses of HTI is essential for the development of the field. The main aspects of human emotional behaviour can offer designers essential information in their search for good design solutions.

Emotions form a central action control system in the human mind and have therefore a focal role in explaining people's behaviour (Ekman 1999; Frijda 1986, 1988, 2007; Oatley et al. 2006). Evolutionarily, emotions constitute a relatively primitive system, and many characteristics of human emotions can be found in other animals as well (Darwin 1872/1998; Panksepp 1998). Emotional processing is mainly centred in the subcortical areas of the human brain, which form a primitive (but in many ways the most fundamental) control system in the mind (MacLean 1990; Luria 1973; Rolls 2000).

Emotions are holistic by nature. In addition to psychological aspects (Ekman 1999; Frijda 1986, 1988; Oatley et al. 2006; Power and Dalgleish 1997), they include important biological dimensions (Panksepp 1998; Rolls 2000), as well as a number of important social features (Eisenberg 2000; Frijda 1986, 1988; Lazarus and Lazarus 1994; Niedenthal et al.

2006). In research it is good to consider the relevance of all aspects when solving design problems.

In the mind, emotions are associated with a moment in time. While people speak, for example, about general happiness, in the mind happiness is always linked to a particular situation. *Emotional state* is the core theoretical concept that expresses how people feel in a certain context of emotional mental representation at a certain time. It illustrates the emotional aspect of a prevailing mental representation. Emotional states are thus aspects of mental states, and they bring with them some aspects of the mental contents of mental representations.

Emotional states consist of a number of basic dimensions. The first is the momentary intensity of emotions: *arousal* (Kahnemann 1973; Power and Dalgleish 1997). There is a graded structure typical to all human emotions, and the intensity or level of arousal is used to express how emotions vary. Irritation, for example, is an emotional state that expresses a negative feeling towards an object, person, or state of affairs (Bono and Vey 2007; Norman 2004). Hate, violence, and rage can be seen as more intense versions of the same emotion. The level of human performance depends on the intensity of the prevailing emotional state (James 1890; Kahnemann 1973). When arousal is very low, as in sleep, the human capacity is also low. When arousal increases, the performance capacity also increases. However, after a certain point, increasing arousal no longer improves the level of performance, and performance starts to deteriorate. When in a panic, for example, people seldom act in a rational manner as their cognitive capacity has decreased. Understanding this principle, called the Yerkes–Dodson law, and other aspects of arousal supports the design of technical artefacts and systems that are meant to be used in critical situations or under pressure (Kahnemann 1973; Salehi et al. 2010). This information implies that critical safety systems should be simple and people should be extensively trained to use them.

Another dimension that concerns emotional states is the length of the duration of an emotion. A person's response to a certain incident may be very short, such as a rapid reaction to a surprise (Power and Dalgleish 1997). Sometimes, in technology design, this kind of surprise-related emotional state is called a *wow effect*. The duration of an emotional state can also be long, and last from weeks to even years, as in the case of different *moods* (Power and Dalgleish 1997). Finally, there are emotional states that belong to one's personality, which represent *attitudes* or *features of personality* rather than affective emotional responses to the surrounding environment (Gross 1998; Niedenthal et al. 2006). A good example of a long-lasting emotional tendency is *temperament*, which is formed in infancy and early childhood and changes only slowly afterwards (Clark and Watson 1999).

When focusing on emotional states in the design, one should also investigate the actual emotional content. Emotional states can differ in their information content. Euphoria, for example, differs from depression in its positive nature, and sadness in its negative nature. Consequently, technical artefacts may include different aspects of emotional content, for example when considering the feelings aroused by the art design of a product. In this kind of analysis, two new dimensions of human emotion become relevant: *emotional valence* illustrates the negativity or positivity of the emotion, and *emotional theme* refers to its general contents (Lazarus 1991; Niedenthal et al. 2006; Oatley et al. 2006; Saariluoma and Jokinen 2014).

The concept of *valence* arises from a pair of opposite emotions—one positive and one negative (Lazarus and Lazarus 1994; Niedenthal et al. 2006; Schmitt and Mees 2000; Spinoza 1675/1955). For example, joy is a positive emotion and sorrow its corresponding negative counterpart. In fact, emotional valence is strongly connected to the pleasantness or unpleasantness of the emotional state, and it determines the desirability of emotional contacts. The positivity of an emotional contact is often seen as imperative in HTI (Jordan 2000; Hassenzahl 2011).

People have many different types of positive and negative emotions, and valence is not the only dimension along which the contents of emotions can be analysed (Lazarus 1991; Power and Dalgleish 1997). The concept of emotional theme—or core emotional theme—is also needed to cover all the aspects of emotional states.

An example may clarify the difference between valence and theme. The feeling of joy in using technologies usually embodies a sense of positive emotions such as wellness, commitment, and positivity (Mayring 2000). Trust is also a positive emotion that, in the context of HTI, refers to human reliance on a given technology. Thus, these two emotions—joy and trust—have the same positive valence but different themes. Hence, the theme defines the accurate contents of emotional states in a more sophisticated manner than valence (Oatley et al. 2006).

Because emotions determine the subjective meaning of a situation to people, they are closely connected to the tendencies of human action (Frijda 1986, 1988; Ortony et al. 1990). They indicate the personal meaning of the target and the possible action that people might carry out (Frijda 1986, 1988; Lazarus and Lazarus 1994). The feeling of fear, for example, makes people flee, whereas curiosity often results in approaching a target. Because these emotional characteristics are built into HTI processes, emotions are a critical element of situational representations.

Emotional responses to things or incidents are not static by nature, and thus do not stay the same forever. Instead, human emotional reactions continue to develop during a person's life span (Oatley et al. 2006; Lazarus 1991; Power and Dalgleish 1997). For instance, people who reacted hastily or aggressively in certain incidents in their youth may behave moderately and calmly in the same situations in maturity, as their emotional make-up has altered. This process of emotional development is called *emotional learning*.

Emotional learning processes change the contents of emotions stored in the memory—so-called emotional schemas—which people use when they select information during perception and build memory representations (Beck 1976; Bower 1981; Williams et al. 1997). Emotional learning is common in HTI. A user who once regarded mobile services as redundant may later become an advocate of this kind of technology after having learned to use and understand its practical value. In this case, a change in an emotional meaning is explained by a change in the contents of emotional schemas that are used by apperception to construct emotional states.

Appraisal: Cognition in Emotions

Emotional processes are closely linked to cognition. An emotional state is not activated by accident; it is based on an individual's understanding of the current situation. If the situation is cognitively seen as risky or threatening, the emotional states are constructed based on danger-related emotions, such as excitement, fear, and courage. If positive relations dominate the situation, emotional states are characterized by relaxation, happiness, humour, and benevolence. Before the situation-related emotional representation is constructed in the human mind its meaning and the cognitive content must be clear to the individual (Frijda 2007).

The psychological process of linking a situation's cognitive and emotional representations is called *appraisal*.[1] Appraisal is a core process in the *psychology of emotions* (Frijda 1986, 1988, 2007; Moors et al. 2013; Weiner 1985; Scherer 2005), which is often defined as the representation of emotional significance for the experiencing person, and the associated emotional value of cognitions and actions (Frijda 1986; Oatley et al. 2006; Ortony et al. 1990). Emotions associated with technologies are relevant in the study of HTI. When performing a similar task, one user may feel anger for failing in the task, while another may feel guilt (Myhill 2003).

The connection of cognitions and emotions is a central issue in user interface design. If people cannot use a technical artefact effectively they become frustrated (Saariluoma and Jokinen 2014), and this activated emotional state is connected to the cognitive characteristics of the given situation. For example, forms and colours may activate cognitive representations, which in turn engender relevant positive or negative emotions (Norman 2004).

The outcome of appraisal is a mental representation with cognitive and emotional dimensions. For example, human cognitive beliefs are connected to their active emotional states. If users do not think they can learn to use e-learning technology they become frustrated (Juutinen and Saariluoma 2007). Thus, cognitive assessment leads to emotional frustration.

These dimensions are not always rationally connected, as people may emotionally represent situations in an inadequate manner and in the case of technology usage, even misrepresent the technologies or their uses both cognitively and emotionally. This may affect the feeling of self-efficacy, that is, an individual's confidence in his or her ability to successfully per-

---

[1] Appraisal can also be seen as an apperceptive process associating emotional values with cognitions.

form a task (Bandura 1977, 1986, 1997). Mistaken beliefs of one's own incapacity may even lead to a self-fulfilling misconception.

Repeated failures create a negative atmosphere and lower self-efficacy, whereas success in using technology improves self-efficacy and generates positive feelings and pride. This makes people more willing to accept, use, and train to use new technologies to achieve their goals (Juutinen and Saariluoma 2007). The example illustrates how appraisal is significant in indicating personal meanings of technologies to people, how people differentiate their emotions, as well as what their behavioural responses are on both the cognitive and physiological levels (Ellsworth and Scherer 2003; Moors et al. 2013). Thus, appraisal connects cognitive evaluation with emotions and actual human actions. Individual preferences and selections of actions are constructed based on representations made in appraisal.

The main question in appraisal-based HTI research is to define what kinds of cognitions activate certain kinds of emotional states. When people interact with a user interface, human mental representation and the respective emotional representation are created. These cognitive and emotional representations control human actions and define whether people like and accept a technology. Thus even a small emotionally mis-placed detail in a user interface can easily decrease the value of the tech-nology in people's minds. One the whole, appraisal can also be seen as an apperception process, as it constructs from its parts the contents of mental representations.

Emotional Aspects of HTI

A number of paradigmatic concepts and discourses have been developed to analyse the emotional aspects of technology usage (Hassenzahl 2011; Helander and Khalid 2006; Nagamashi 2011; Norman 2004; Jordan 2000). Some of them discuss emotional features in products that activate users' emotional states, while others are interested in what takes place in users' minds and what the main characteristics of technology-relevant emotional states are.

A classic example of emotion-based thinking in technology design is Kansei engineering, which applies traditional psycho-semantic methods to analyse emotional aspects in different technological and product contexts (Nagamashi 2011). There are also other paradigms such as emotional and affective usability research (Helander and Khalid 2006; Norman 1999, 2004), user experience research (Hassenzahl 2011; Kuniavsky 2003; Norman 1999), affective design, pleasurable design (Jordan 2000), and funology (Monk et al. 2002). All of these are relevant approaches in investigating emotional interaction. These closely related scientific paradigms have illustrated the importance of emotional dynamics in technology design thinking (Helander and Khalid 2006; Kuniavsky 2003).

The relevance of emotional interaction design is a recognized challenge among designers (Brave and Nass 2009; Hassenzahl 2011; Norman 2004), the significance of which can be seen in many design outcomes of information technology. A great deal of effort is put into designing products with faddish and distinguishable forms and symbolic values (Nagamashi 2011). This trend can be clearly seen in consumer products, and even robots are designed with more and more humanized elements, such as warm smiles and pleasant voices.

In traditional industrial design, when creating brands, designers bring out certain product characteristics to create particular emotional states (Stenros 2005). Scandinavian design, for example, is known for its minimalist, functional, and practical forms that reflect people's relationship with nature and the wild (Englund and Schmidt 2013). These forms can be found in the works of such famous designers as Arne Jacobsen, Poul Henningsen, and Alvar Aalto. The pursuit of the ideals of functionality and practicality can be seen in the designs of information technology, too. Many devices are meant to represent the feeling of simplicity and effectiveness, at least from the usability point of view.

Emotional interaction does not, however, concern only the immediate use of a product or system. Technologies should also respond to emotions such as trust and confidence, which also reflect people's attitudes towards technology in general. People have to trust that the given technology operates as promised, and that it will not create or exacerbate any practical or ethical problems. For example, in order to efficiently serve citizens in society, technical applications and systems increasingly collect private information about people (e.g., personal codes and account information). Citizens have to be able to trust that this information is well protected, and that no one can use maliciously. In the case of smart cards for example, users need to be confident that the system will reliably and correctly identify them and not permit access to any other users.

Negative trust in technology is embodied in such emotional states as technophobia (Brosnan 2002), the dislike of technology. When confronting new technology, people may feel incapable of using it. Novice and elderly users often deem poor interaction to be the consequence of their own inability to use a given technology. At workplaces, poor usability may even lead to psychosomatic stress among people who feel that they have been coerced into working with systems they cannot handle (Arnetz and Wiholm 1997). Technophobia can disappear as a consequence of direct (Kelley and Charness 1995) or indirect (Rogers and Fisk 2000) positive experiences and improving skills (Ellis and Allaire 1999).

The concept of trust illustrates the connection between emotions and cognitions. In the context of appraisal, when people trust each other or a technology, emotions and cognitions are connected. On a cognitive level, a person may be aware that the information security system installed on her computer is powerful enough to prevent all virus attacks. The reasons for believing this are based on, for example, previous experiences with online services and information offered by computer support services and the media. This information constitutes the knowledge base that creates a feeling of trust. In an immediate usage situation (using a banking system, for example), the feeling of trust is realized as determined input actions when operating the user interface. In a broader context outside the immediate use of the system, trust defines the overall attitude towards banking systems, for example at the moment of purchase.

Emotions are present every minute of the day, and their influence is widespread in HTI as well. For example, an extreme case of the influence of emotions in Internet dependency—the compulsory and excessive use of the Internet in the form of computer games or social media (Block 2008). The addiction develops as a result of the short-term pleasure enjoyed when playing games or surfing the net (i.e., when the device is turned on). Internet dependency is classified as a psychiatric illness, which requires professional intervention. In South Korea alone, it has been assessed as one of the most serious public health issues, as it has been claimed to be connected to, for example, childhood obesity. The ever-increasing time spent in front of the Internet or a computer game change eating habits, increasing the number of snacks and the amount of unhealthy food consumed. Also, when spending many hours a day on the Internet, young people are constantly tired and unable to concentrate in school, as a result of which they fail their exams.

On the grounds of knowledge of the nature of people's relevant emotional states, it would be possible to generate rational design goals for technologies, for example to assess aesthetic requirements for the product and to look for emotionally inspiring and satisfactory ways of using technologies. Recent developments to encourage personal health monitoring systems are a good example. The field of emotional user psychology opens up an extensive set of design questions and challenges, of which only a small percentage has yet been investigated.

User Motivation: Biological, Cognitive, Emotional, and Social

In addition to emotions, dynamic psychology is concerned with human motivation, which in psychology refers to the mental mechanism or forces driving, maintaining and controlling human goal-directed actions (Atkinson 1964; Cofer and Appley 1968; Franken 2002; Laming 2008)—that is, the *conative* dimension of the mind (Hildgard 1980). Motivation is thus relevant in explaining why people use technologies and in understanding to what extent different design solutions help motivate people to use particular products and systems. As mentioned earlier, motivation refers to people's energy to pursue their goals. In the context of HTI, one can speak about user motivation.

Examining motives provides answers to the question of why people carry out different tasks. Motivation research focuses on people's goals and how energetic their pursuit of these goals is. People join e-courses to acquire different skills, from which they could benefit later in life. In this example, the goal state of acquiring knowledge for the future forms the primary motivational aspect, and attending a course forms the secondary motivational aspect.

Research on users' motives can offer fundamental knowledge to support design decisions. One of the most important focus areas in this context is the introduction of new technology (Dvash and Mannheim 2001; Shelton et al. 2002; Venkatesh 2000). In the HTI literature, an outstanding example of technology introduction is the model of *technology acceptance* (TAM) (Davis 1989; Venkatesh 2000). The basic idea is that perceived usefulness and ease of use are crucial factors in explaining how motivated people are to accept a given technology. 'Perceived' (or conceived) refers to the cognitive representation of one's ability to use a technology.

Motivation has biological, cognitive, emotional, and social elements, and it opens up a new independent discourse in research into the human mind (Franken 2002). Understanding motivation is central in analysing why a defined goal is important for a certain group of people. For example, the motivation to use a product or system is present throughout the action that is carried out. Thus the motive is not a cause for action; rather, it is the action itself. Therefore, designing how to maintain motivation is as important as designing how to spark the initial motivation to use something. The Rubik's cube was a global success in its time, but it—like many other products—has lost much of its popularity since. To avoid a loss of interest among users, the motivation to use technology should be more rigorously examined as part of the lifespan design of products and systems.

Human needs form the basis of human motivation. People generally pursue the satisfaction of their needs (Maslow 1954).[2] Different theories of need hierarchy, such as Maslow's (1954) well-known model, have been presented in the literature of ergonomics and interaction. Maslow's (1954) idea was that some human needs are more basic than others, and that only after satisfying such basic, lower-level needs such as thirst,

---

[2] It is important to understand the distinction between the two concepts: engineering interaction books routinely entail the concept of '*user need*', which is an intuitive concept that has only an indirect connection to psychological needs. Thus it is important to separate 'user need' from human needs. The purpose of technology is to have a function, often related to very complex systems of human life, and therefore it should not be simplified to such overall concepts as user need.

hunger, and sex is one able to proceed to higher-level needs, such as self-actualization. The important dimension of Maslow's work is his concrete introduction to the connection between needs and motives, although modern research on motivation has illustrated that it cannot be conceptualized solely in terms of static need hierarchies.

The basic structure of needs and motivation has been described by applying the concept of the depletion–repletion cycle (Toates 1986). In this description, human physiological needs activate people towards certain goals, which enable them to satisfy their needs. The state before satisfaction can be called *desire*, as it has a goal but not necessarily the means to reach it. When a person finds a way to satisfy a need, the need state is deactivated or depleted. When the need has been satisfied, the action can be shifted to another direction. In everyday life, the mechanisms of satisfying needs can be complicated. People may save money to use later in life without having any clear plan about how they intend to use it, or what kinds of needs it would satisfy.

Biological psychology and *neuroscience* can be illuminating in the analysis of motives (Berridge 2004). Likewise, neural phenomena can offer the possibility of objectively understanding many aspects of motivation when people use technologies, although biological concepts and theories alone cannot explain human motivation in technology usage. This kind of research can be carried out in the field of neuroergonomics (Parasuraman and Rizzo 2006), which uses biopsychological techniques to analyse needs and motives. For example, an individual's heart rate can be used to analyse stress during technology use (Anttonen and Surakka 2007).

Emotions also have a central role in understanding motivation, as they inform people about needs and other motivational states. The concept of pleasure, for example, is usually seen as a motivational goal towards a certain state. People mostly pursue pleasurable states, and products that help them achieve this state are considered favourable and motivating (Hassenzahl 2011; Jordan 2000). For example, the need for water evokes thirst, which makes people start to look for a drink. Thus the concept of pleasure, in addition to being an important goal, is also an emotional state. Depending on the situation, there can of course be numerous other emotional states that explain the motives behind certain actions.

Engagement is another example of motivating emotional states (Porges 2003). In the HTI context, engagement refers to a commitment to a product or brand. In sports, for example, engagement to certain products is obvious (Dickey 2005). Engagement involves long-lasting motivation, and explains why people choose some products rather than others.

An additional perspective on motivation in HTI is revealed by the differences between motives connected to self-based emotional states. Motivation is a combination of two types of factors influencing human actions: *extrinsic* and *intrinsic* motives (Deci et al. 1999). Extrinsic motives involve different external factors that influence human motivation. For example, sports exercise can be accomplished for extrinsic reasons such as improved fitness or better appearance. Intrinsic motives, on the other hand, are internal to the individual. In the case of sports exercise, motives such as enjoyment and competence can be intrinsic. In sports, intrinsic motives remain a more critical factor in sustained physical activity, whereas extrinsic motives often fail to sustain exercise over time. A person cannot always directly influence extrinsic motives. For example, employers mostly decide on salary and other organizational incentives, and employees have little ability to influence them. The motivation to use technologies can also be extrinsic, so that people use technical artefacts because their employer expects them to.

Intrinsic motives are internal, and people set these motives for themselves. For example, the desire to express oneself is an important motivating force in the early adoption of new technologies (Kim et al. 2001). Students, for example, are more motivated to put effort into learning to use technology because of the challenges of the tasks, rather than because of some course rewards. Similarly, playful and emotionally positive interaction with a computer may be motivating. For example, a computer game should motivate players intrinsically if they are to be expected to continue with it for any length of time (Venkatesh 2000).

Intrinsic motivation is a valuable asset in marketing any product. It includes several forms, such as self-actualization and self-determination. Maslow (1954) argued that self-actualization has a vital role in the human control of actions (May 2009). Later, Bandura (1986) and Ryan and Deci (2000) demonstrated that self-determination and self-efficacy can explain how intrinsic motives are more dominant than extrinsic motives.

The former concept refers to how people assess their capacity to cope with a task and the latter to how competence, relatedness, and autonomy form the core of intrinsic motivation.

Self-determination is a valuable motive when introducing new technologies to people. People must have the right to decide what kind of technology they want to use. Although this may not always be possible, forcing consumers to use new service systems, for example, easily leads to rejection and negative attitudes towards the service provider in question. Without a right to decide, extrinsic motivation becomes dominant and may weaken the likelihood of successful technology adoption (Reinders et al. 2008).

Human motives cannot only be explained on emotional grounds, for they also have *cognitive* dimensions. Perhaps the best-known link between cognition and motivation is *cognitive dissonance*, which refers to a motivational state dominated by conflicting cognitions. For example, people may want to buy a new technical device, but because it is said to have been produced unethically, they may be afraid of losing face by buying it. Other people may have purchased a product but feel uncertain about the wisdom of their decision (Kahnemann 2011).

Cognitive aspects of motivation can be associated with the scope that technology can offer in improving users' tasks. In organizations, for example, people are willing to use new technologies if they understand how these new tools will improve upon their earlier practices (Orlikowski 2000). Yet poor understanding of the outcome of technology use may cause opposition. Such cognitive phenomena as intention, learning, perceived self-efficacy, and usefulness are factors that explain the cognitive aspects of motivation when using technologies (Yi and Hwang 2003).

Finally, there is a *social* aspect of motivation (Dunning 2004; Ryan and Deci 2000). In fact, numerous motives in human life are social by nature. Social dimensions of technology use can often explain why people are motivated, or why they are less motivated when using certain technical artefacts. These themes are focal today thanks to computer-supported cooperative work, and especially to social media applications such as blogs, forums, wikis, and social networks (Tapscott and Williams 2008).

Social media has important functions in the formation of modern social motives. Social motives become visible in goal setting, goal contents, and the choices made between different goals (Franken 2002; Karniol and Ross 1996). This is an important phenomenon, for example, in creating brands and organizing people around them. Typical examples are communities around Apple or Linux (Himanen 2010). On this level, motives can be formulated as goal-related beliefs, and emotions as either congruent or incongruent with establishing goals (Franken 2002).

Because people use technologies to support their actions, motivation is a central field in the investigation of human–technology relations. The motivation to use technologies (and the resulting success rate) relies to a large extent on human *self-image* (Santos-Pinto and Sobel 2005). If one does not believe in one's competence to use a technology, the positive learning cycle comes to a stop.

In HTI literature to date, relatively little interest has been paid to motivation compared to discussions about such matters as usability. However, as can be seen from the analysis above, when developing HTI research and design, it is essential to put more effort into connecting general knowledge about human motivation with specific and important HTI problems.

Personality, Technology Generation, and Interaction Dynamics

Personality is the last concept to be discussed in the context of dynamic psychology. It has a central role in investigating HTI. The psychology of personality studies human individuals as integrated wholes (Mischel 1993). It also focuses on the cognitive, emotional, and conative (or motivational) dimensions of the mind. The psychology of personality asks what a person is or what a person is like, and identifies similarities and dissimilarities between individuals (Hogan et al. 2000; Mischel 1993; Pervin and Oliver 1997). It also concentrates on stable response patterns that individuals have in different types of situations. In modern psychological terms, this refers to the stable content of individuals' internal representations.

The core questions of the psychology of personality have great significance in user psychological interaction analyses. Human personality explains how people as a whole consider technologies. Such elements as dispositions, attitudes, habits, appeals, beliefs, values, and aspirations, for example, vary depending on the person—and even within an individual over longer periods of time. Hence, answers to many questions essential to user research can be based on the concepts and methods of the psychology of personality. This kind of knowledge enables one to segment and characterize target user groups, as well as study what kind of technical artefacts they would need and accept.

The definition of different user groups has become increasingly important in recent interaction design. In addition to making devices that people *can* use, attention is paid to the *characteristics* of individuals in the target group by analysing personality features and other differences using factorial methods (Hogan et al. 2000; Mischel 1973). A well-known example is the so-called Big-5 theory, which focuses on five basic personality traits: extraversion, agreeableness (or warmth), conscientiousness, stability, and openness to experience (or intelligence) (see, e.g., McCrae and Costa 1987; Wiggins 1996). In Big-5 analysis, based on their answers to test questions, people are placed differently in this five-dimensional space.

Knowledge of personality types can be used to target products to different types of consumers. Users' or consumers' commitment to a product may depend on different aspects of their personalities (Kim et al. 2001). Some product features may appeal to many users because they serve as status indicators and have symbolic value (Caprara et al. 2001; Fournier 1998). Analysis of these kinds of factors that connect the notions of personality and the user is still in its infancy, but it is self-evident that personality is a key concept in considering how people as whole accept and use technologies.

An individual's personality also reflects a wide range of possible mental contexts, including attitudes and values. *Attitudes* are acquired tendencies to apperceive situations in a certain way. For instance, Linux specialists may regard their own operating system as different from that of a normal microcomputer user (Himanen 2010; Moody 2002). *Values* are an important closely related concept or type of attitude that indicate what a person considers to be good (e.g., Runes 1983).

Attitudes and values can also be seen as individual and personality factors. They are relatively stable over a person's lifetime. Therefore, they have a fundamental role in how people experience reality, receive information and make decisions (Aiken 2002), and they also hold a central position in consumer psychology (Jacoby et al. 1998) and the closely related field of consumer psychology.

The formative period in life, the period of adolescence and young adulthood, is critical in terms of attitudes towards future technology. During this period, cognitive abilities and experiences influence each other for further maturation. Accordingly, the understanding of how to use technology (present and future) is built on the kind of knowledge that is typical for that cohort as a technology generation (Docampo Rama 2001; Bouma et al. 2009). The concept of technology generations is thus built on the idea that technology that is available in one's formative period creates practices and mental models that will later affect how people encode new technology. From this point of view, designers have to be aware of the potential generation-specific ability gap between users when creating new types of user interfaces and interaction models. Appreciating this generational difference does not only concern the design of tangible user interfaces but also understanding such issues as privacy and trust, for instance, in the usage of the technology in question (Leikas et al. 2011).

Attitudes form an interesting group of phenomena in analysing human mentality (Osborne et al. 2003). Their stability makes it possible to use them to explain many different types of phenomena in HTI (Aiken 2002; Zimbardo and Leippe 1991), and to forecast certain forms of human behaviour rather accurately.

Attitudes are activated automatically, and they control actions; they are stable, but not immutable (Aiken 2002; Osborne et al. 2003; Zimbardo and Leippe 1991). Attitude is often a system of beliefs linked to emotional models that govern the selection of information, in the sense that information that is contrary to prevailing attitudes is rejected, and information that is positive is adopted (Aiken 2002). A special branch called attitude psychology can help solve problems caused by users' expectations. For example, company image is dependent on customers' attitudes.

A negative image swiftly becomes expensive even for the most powerful companies (Teo 2002).

Change in attitudes is one of the core questions of user psychology. New functions in the ICT field are often adopted relatively slowly. For example, information networks existed long before they became widely used in the 1990s, and mobile phones were also quite slowly adopted for everyday use. Today one of the field's greatest challenges is interesting users in new mobile services. These problems are not only technical; the accompanying attitudes also set important challenges for designers.

Cultures and Their Differences

People as individuals are always surrounded by numerous groups of other people, the largest of which can be called cultures: ways of life that are shared by groups of people. In the contexts of cultures, people reach their adulthood and internalize different norms and values. The mental contents of a culture thus become parts of the mental contents of people living in this culture (Matsumoto 2000). HTI design must consider that the development of ICT is a global phenomenon that affects different cultures in different ways.

Culture can be seen as an integrated system of customs, symbols, languages or dialects, values, and beliefs (Geertz 1973; Harris 1979). The system can be national, organizational, religious or symbolic, or even virtual. In different cultures people may differ essentially from each other; they have their own myths and legends, and different ways of solving social problems, for instance.

Cultures have their own institutional ways of solving production problems and defining the position of individuals (Castells 1997). They are often seen as geographic and even national systems. However, not all cultural issues are localizable. Organizations have their cultures, young and old people have their cultures, and ethnic groups, male and female as well as people with specific skills such as experts all have their own cultures (Hofstede 1980). Therefore people do not belong to only one kind of culture at a time. They belong to groups, sub-cultures (such as teens), or organizational cultures (Hofstede 1980) and follow the values and ways of behaving that are typical to that culture; people of the same culture often share mental representations and underlying patterns of thought. Characteristics that are significant in one culture are often less important in others, which creates challenges for HTI design.

For example, cultures often have different approaches to communication and semiotics such as cultural symbols, terms, and even the meaning of colours. Three principles are particularly useful when designing for cultural diversity (Matsumoto 2000). First, it is important to understand the main principles and findings of cross-cultural psychology. Second, one has to understand other cultures and their internal logic. Finally, it is essential to make realistic case-specific conclusions and avoid stereotypical thinking.

The notion of culture will eventually change as people become increasingly networked, creating more or less global cultures (Castells 1997) in which an endless number of people share an interest and form associated social groups. ICT has already changed cultures with its capacity to connect people in new ways (Castells 1997). Social media has created numerous sub-cultures and joined people globally with a similar mindset. Geographical location, though not meaningless, is no longer as important as it used to be (Castells 1997). Anticipating cultural and lifestyle changes is essential when designing for the future.

The Dynamic Nature of the Human Mind

This chapter describes how the human mind always operates in a holistic manner. Considering dynamic HTI and users' minds, cognitive, emotional, motivational, and personality phenomena are always present and linked together. Cognitions influence emotions, while emotions influence motives and personality sets preconditions on all of them. Personality also connects emotions and cognitions to such wider issues as social groups and culture. In science, it is normal to set aside other dimensions and only concentrate on one specific issue of a larger phenomenon, but in technology design it would be impossible to create a successful outcome without taking into account all the above aspects of HTI.

The theoretical concept of *mental representation* is useful for considering cognitive and dynamic problems (Markman 1999; Newell and Simon 1972). Mental representations always have both cognitive and emotional aspects, both of which are central in explaining motives and attitudes. They can have cognitive contents such as perceptions, beliefs and thoughts, but they also have emotional dimensions, such as amiability, anger, or envy. Finally, the contents of human mental representations depend on one's personal life history and other personal characteristics.

Research into the dynamic aspects of interaction offers new perspectives in interaction design. In many design challenges, for example when designing new media services, it is helpful to have knowledge about users as personas. Dynamic processes direct and maintain people's activities, and can thus explain and clarify the solutions to many important design problems (cf. Ambrose and Kulik 1999; Carlson et al. 2000; Greenberg and Baron 2000).

The overview of dynamic interaction presented in this chapter is not meant to be exhaustive; it is intended to introduce the versatility and scope of the problems of dynamic interaction in order to highlight the challenges of dynamic user psychology and manage design thinking in product development.

User experience can be used to illustrate the HTI phenomenon in terms of mental representations. An increasing number of literatures on different aspects of user experience are produced (Norman 2004; Law et al. 2008), but the concept is still rather ambiguous and elusive, thus far lacking sufficient empirical research or an exact definition (Hassenzahl 2011; Hassenzahl and Tractinsky 2006). However, experiences are conscious cognitive and emotional states of mind, and thus the concept of user experience is relevant when considering the connections between dynamic and cognitive aspects of the user's mind during interactions. Like mental representations, experiences also have cognitive and emotional contents. *Cognitive contents* include perceptions, remembrances, and thoughts. Understanding why people are motivated to use certain technologies requires knowledge of the *emotional and motivational contents* of their experiences.

Motives and emotions have multifaceted roles in actions (Deci et al. 1999; Franken 2002; Maslow 1954; Ryan and Deci 2000; Switzer and

Sniezek 1991). The interconnectedness of needs, emotions, and motives is clear in various motive phenomena based on rewards and punishments (Franken 2002). The satisfaction of needs also affects emotional states: states with a negative valence are usually punitive, and those with a positive valence are rewarding. Naturally, rewards encourage efforts towards a goal, while punishments discourage particular actions. For the user, the fundamental reward is often a positive user experience. This motivates and directs people to fulfil their needs in the context (of use) in question. An obscure service is not encouraging, and only an external imperative will make people use a low-quality service. Even if a user receives training to use a poor-quality service, it may in any case be quickly forgotten.

In addition to cognitive and emotional concepts, experience can also be considered the conscious part of mental representations, which also include unconscious elements (Allport 1980; Dienes and Scott 2005; Ellenberg 2008; Freud 1917/2000; Marcel 1983). Because the conscious mind is just the tip of the iceberg, understanding experience—including user experience—also requires understanding the unconscious elements of representations, which cannot be verbally expressed (Saariluoma and Jokinen 2014).

Motives and emotions have multifaceted roles in actions (Deci et al. 1999; Franken 2002; Maslow 1954; Ryan and Deci 2000; Switzer and Sniezek 1991). The interconnectedness of needs, emotions, and motives is clear in various motive phenomena based on rewards and punishments (Franken 2002). The satisfaction of needs also affects emotional states. States with a negative valence are usually punitive and those with a positive valence are rewarding. Naturally, rewards encourage efforts towards a goal while punishments discourage this. For the user, the fundamental reward is often a positive user experience. This motivates and directs people to fulfil their needs in the context (of use) in question. An obscure service is not encouraging and only an external imperative will make people use a low-quality service. Even if a user receives training to use a poor-quality service, the service may in any case be quickly forgotten.

From the user's perspective, the interaction between the user and the technical artefact is not only collaboration between human senses and limbs; emotions and motives—linked to perceptions, attention, values, and memory—also play a significant role. The *emotional interface* is a fundamental part of human experience, and therefore the emotional design of usability also involves analysing the user's cognitive experience. Cognitions activate emotions by cognitively assessing the situation, and emotions control cognitive processing. Consequently, the origins of the interaction phenomena must be uncovered in each case (on emotions and cognitions, see, for example, Beck 1976; Wells and Mathews 1994).

The cognitive mind selects which emotions will be activated (e.g., fear or delight), and these emotions create motives for action (Lazarus and Lazarus 1994; Power and Dalgleish 1997). As a consequence, respective action tendencies are activated, and motivated actions are carried out. People cognitively define the interaction situations from which they can construct their dynamic interpretations. If the perceptions are incorrect, the emotional responses may be inappropriate as well. However, correct perceptions facilitate the emergence of correct emotional interpretations.

People identify the emotionally important features in the environment through an emotional assessment process. For example, distressed test subjects may process words that refer to anxiety more quickly than words that are neutral (see reviews in Wells and Mathews 1994). These findings imply that emotions unconsciously affect the direction of attention. Strong emotional intensity also has a tendency to narrow the attention (Kahnemann 1973; Wells and Mathews 1994).

Mental dynamics is based on the regular and individual emotional models that people have learned in their lives. This emotional theme controls how aggregate functions are formed. Predilection, for example, is a social emotion in which the person gains a positive emotional experience from the object's happiness (Mees and Rohde-Höft 2000; Power and Dalgleish 1997), whereas feelings of inferiority may make people place unnecessary limitations on themselves (Adler 1929/1997; Beck 1976; Sanders and Wills 2005).

The interaction of cognitions, emotions, and motives is based on the fact that cognitive and dynamic schemas form entities. This connection emerges quite clearly when examining memory processes. Memory can be considered a network of information or a system of schemas in which the content of the working memory is formed by the active nodes of each situation. The nodes of the network represent both cognitive memories and the emotions connected to them (Bower 1981). Cognitive and dynamic states are connected in active representations. They work simultaneously, but the individual's general impression about the world always depends on both cognition and emotion.

It is essential for the psychology of memory that unconscious emotional and motivational schemas often control our activity (Beck 1976; Berry and Dienes 1993). Coca-Cola provides an interesting example of this. When competing with Pepsi in the USA, it put New Coke on the market and withdrew the traditional Classic Coke. As a consequence, the company lost much of its market share as consumers felt that Classic Coke had lost its position as one of the symbols of the American way of life (Tibballs 1999). The example reveals the influence of unspoken emotional representations in how people encode information in their environment. Hidden intuitions may be genuine, but they may also be incorrect. In many cases it is impossible to know if the hidden knowledge is correct (Nonaka and Takeuchi 1994; Saariluoma 1997).

Emotions and their respective motives ultimately define what consumers consider to be good or bad. Thus, they create a direction for attitudes and control various decisions. They also have an essential personal meaning that is manifested in what things people find positive and negative, valuable and not valuable. By learning to understand emotional processes, one can learn to see what kinds of things create the most fundamental representational contents for people.

Values are closely connected to cognitions, emotions, and motives. Values can be moral, practical, or aesthetic. One can offend someone's values and feelings without noticing it, or take the values of others into consideration, for example, by putting oneself mentally in the position of another person to imagine how one would feel in a similar situation and deal with similar incidents. Devices also have various emotional and product values that are linked to them. From the user psychological point of view, it is essential that the designed technology respond to the values of its users, as the willingness to adopt technologies may depend on these values. Therefore, human values should serve as a guideline for designing technology and the information society in general.

Motives provide information about the goals and energy of actions (Franken 2002). Thus, analysing motives can offer concrete information to the analysis of why some people adopt certain technologies. To acquire knowledge of what people expect of a product (and why they would use it) presupposes, in addition to motives, clarifying possible dissonances.

Finally, *personality research* can introduce a new set of questions that allow an inspection of the relationship between technology and individuals with different emotional, cognitive, and motivational patterns. Understanding what kind of people with what kind of motives will be interested in different design makes it possible to segment people and target products to these segments. Thus, the issue of user personality integrates all elements of cognitive and dynamic interaction research.

As the human mind and human action can be seen as a combination of cognitive, emotional, and dynamic (or conative) processes (Hildgard 1980), a part of the discussion has been targeted at three main sub-fields: emotions, motives, and personality issues. It is also vital to consider the joint issues between the fields—that is, how cognitions influence motives—and to assess to what extent personality is a factor in forming cognitions, emotions, and motives.

In all user research, it is possible to begin with an analysis of cognitions relevant to the usage situation, for example by using such methods as thinking aloud, interviews, and surveys (Ericsson and Simon 1984). Second, it is essential to investigate different emotions that are involved in the interaction with the technology in question. If the interaction is long, or includes complicated multistage processes, different usage situations may require separate analysis. Basic psychology provides information about interpreting the results, in terms of what kinds of visual features or which technical details should be altered in searching for solutions that enable people to reach positive emotional states.

The research on immediate interaction situations raises questions such as why a user interface should be experienced in a negative way, why users have difficulties in discriminating between the target and the background or why the user interface should be non-intuitive. It may also be that the user interface does not address cross-cultural demands. These examples highlight the importance of understanding the immediate dynamic interaction between people and artefacts. Since human mental processes are integrated into a whole, it would be impossible to eliminate the sense of frustration in operating a user interface without understanding its psychological source. The frustration might be caused by a number of factors, such as the location of the user interface components on the screen, a lack of cultural knowledge, or a lack of computational skills.

References

Adler, A. (1929/1997). *Neurosen: Fallgeschichten*. Frankfurth am Main: Fischer.
Aiken, L. R. (2002). *Attitudes and related psychosocial constructs: Theories, assessment, and research*. Thousand Oaks, CA: Sage.
Allport, D. A. (1980). Patterns and actions: Cognitive mechanisms are content specific. In G. Claxton (Ed.), *Cognitive psychology: New directions* (pp. 26–64). London: Routledge and Kegan Paul.
Ambrose, M. L., & Kulik, C. T. (1999). Old friends, new faces: Motivation research in the 1990s. *Journal of Management, 25*, 231–292.
Anttonen, J., & Surakka, V. (2007). Music, heart rate, and emotions in the context of stimulating technologies. In *Affective Computing and Intelligent Interaction: Proceedings of the Second International Conference, ACII, 12–14 September, Lisbon* (pp. 290–301). Berlin: Springer.
Arnetz, B. B., & Wiholm, C. (1997). Technological stress: Psychophysiological symptoms in modern offices. *Journal of Psychosomatic Research, 43*, 35–42.
Atkinson, J. (1964). *An introduction to motivation*. Princeton, NJ: Van Nostradam.
Bandura, A. (1977). Self-efficacy: Toward a unifying theory of behavioral change. *Psychological Review, 84*, 191–215.
Bandura, A. (1986). *Social foundations of thought and action: A social cognitive theory*. Englewood Cliffs, NJ: Prentice-Hall.
Bandura, A. (1997). *Self-efficacy: The exercise of self-control*. New York: Freeman.
Beck, A. (1976). *Cognitive therapy of emotional disorders*. Harmondsworth: Penguin Books.
Berridge, K. C. (2004). Motivation concepts in behavioral neuroscience. *Physiology and Behavior, 81*, 179–209.
Berry, D. C., & Dienes, Z. (1993). *Implicit learning: Theoretical and empirical issues*. Hillsdale, NJ: Erlbaum.
Block, J. (2008). Issues for DSM-V: Internet addiction. *American Journal of Psychiatry, 165*, 306–307.
Bono, J. E., & Vey, M. A. (2007). Personality and emotional performance: Extraversion, neuroticism, and self-monitoring. *Journal of Occupational Health Psychology, 12*, 177–192.

Bouma, H., Fozard, J. L., & van Bronswijk, J. E. M. H. (2009). Gerontechnology as a field of endeavour. *Gerontechnology, 8*, 68–75.

Bower, G. H. (1981). Mood and memory. *American Psychologist, 36*, 129–148.

Brave, S., & Nass, C. (2009). Emotion in HCI. In A. Sears & J. A. Jacko (Eds.), *Human-computer interaction: Fundamentals* (pp. 53–68). Boca Raton, FL: CRC Press.

Brosnan, M. J. (2002). *Technophobia: The psychological impact of information technology.* London: Routledge.

Caprara, G. V., Barbaranelli, C., & Guido, G. (2001). Brand personality: How to make the metaphor fit? *Journal of Economic Psychology, 22*, 377–395.

Carlson, N. R., Buskist, W., & Martin, G. N. (2000). *Psychology: The science of behaviour.* Harlow: Allyn and Bacon.

Castells, M. (1997). *The network society.* Oxford: Blackwell.

Clark, L. A., & Watson, D. (1999). Temperament: A new paradigm for trait psychology. In L. A. Pervin & O. P. John (Eds.), *Handbook of personality: Theory and research* (2nd ed., pp. 399–423). New York: Guilford Press.

Cofer, C. N., & Appley, M. H. (1968). *Motivation: Theory and practice.* New York: Wiley.

Damasio, A. (2005). *Descartes' error: Emotion, reason, and the human brain.* Harmondsworth: Penguin Books.

Darwin, C. (1872/1999). *The expression of the emotions in man and animal.* London: Fontana Press.

Davis, F. D. (1989). Perceived usefulness, perceived ease of use, and user acceptance of information technology. *MIS Quarterly, 13*, 319–340.

Deci, E. L., Koestner, R., & Ryan, R. M. (1999). A meta-analytic review of experiments examining the effects of extrinsic rewards on intrinsic motivation. *Psychological Bulletin, 125*, 627–668.

Dickey, M. D. (2005). Engaging by design: How engagement strategies in popular computer and video games can inform instructional design. *Educational Technology Research and Development, 53*, 67–83.

Dienes, Z., & Scott, R. (2005). Measuring unconscious knowledge: Distinguishing structural knowledge and judgment knowledge. *Psychological Research, 69*, 338–351.

Docampo Rama, M. (2001). *Technology generations—Handling complex user interfaces.* Eindhoven: University of Eindhoven.

Dunning, D. (2004). On motives underlying social cognition. In M. B. Brewer & M. Hewstone (Eds.), *Emotion and motivation* (pp. 137–164). Malden, MA: Blackwell.

Dvash, A., & Mannheim, B. (2001). Technological coupling, job characteristics and operators' well-being as moderated by desirability of control. *Behaviour and Information Technology, 20*, 225–236.

Eisenberg, N. (2000). Emotion, regulation, and moral development. *Annual Review of Psychology, 51*, 665–697.

Ekman, P. (1999). Basic emotions. In T. Dalgleish & M. Power (Eds.), *Handbook of cognition and emotion*. Chichester: Wiley.

Ellenberg, H. (2008). *The discovery of unconsciousness*. New York: Basic Books.

Ellis, R. D., & Allaire, J. C. (1999). Modeling computer interest in older adults: The role of age, education, computer knowledge, and computer anxiety. *Human Factors: The Journal of the Human Factors and Ergonomics Society, 41*, 345–355.

Ellsworth, P. C., & Scherer, K. R. (2003). Appraisal processes in emotion. In R. J. Davidson, K. R. Scherer, & H. Goldsmith (Eds.), *Handbook of affective sciences* (pp. 572–595). New York: Oxford University Press.

Englund, M., & Schmidt, C. (2013). *Scandinavian modern*. London: Ryland Peters & Small.

Ericsson, K. A., & Simon, H. A. (1984). *Protocol analysis*. Cambridge, MA: MIT Press.

Fournier, S. (1998). Consumers and their brands: Developing relationship theory in consumer research. *Journal of Consumer Research, 24*, 343–353.

Franken, R. (2002). *Human motivation*. Belmont, CA: Wadsworth.

Freud, S. (1917/2000). *Vorlesungen zur Einführung in die Psychoanalyse*. Frankfurth am Main: Fischer.

Friedman, S. M. (2011). Three Mile Island, Chernobyl, and Fukushima: An analysis of traditional and new media coverage of nuclear accidents and radiation. *Bulletin of the Atomic Scientists, 67*, 55–65.

Frijda, N. H. (1986). *The emotions*. Cambridge: Cambridge University Press.

Frijda, N. H. (1988). The laws of emotion. *American Psychologist, 43*, 349–358.

Frijda, N. H. (2007). *The laws of emotion*. Mahwah, NJ: Erlbaum.

Geertz, C. (1973). *The interpretation of cultures: Selected essays*. New York: Basic Books.

Greenberg, J., & Baron, R. A. (2000). *Behaviour in organisations*. Upper Saddle River, NJ: Prentice-Hall.

Gross, J. J. (1998). The emerging field of emotion regulation: An integrative review. *Review of General Psychology, 2*, 271–299.

Harris, M. (1979). *Cultural materialism*. New York: Random House.

Hassenzahl, M. (2011). *Experience design*. San Rafael, CA: Morgan & Claypool.

Hassenzahl, M., & Tractinsky, N. (2006). User experience—A research agenda. *Behaviour and Information Technology, 25*, 91–97.

Helander, M., & Khalid, H. M. (2006). Affective and pleasurable design. In G. Salvendy (Ed.), *Handbook of human factors and ergonomics* (pp. 543–572). Hoboken, NJ: Wiley.

Hildgard, E. (1980). Consciousness in contemporary psychology. *Annual Review of Psychology, 31*, 1–26.

Himanen, P. (2010). *The hacker ethic*. New York: Random House.

Hofstede, G. (1980). Culture and organizations. *International Studies of Management and Organization, 10*, 15–41.

Hogan, R., Harkness, A., & Lubinski, D. (2000). Personality and individual differences. In K. Pawlik & M. R. Rosenzweig (Eds.), *International handbook of psychology* (pp. 283–304). London: Sage.

Jacoby, J., Johar, G. V., & Morrin, M. (1998). Consumer behavior: A quadrennium. *Annual Review of Psychology, 49*, 319–344.

James, W. (1890). *The principles of psychology*. New York: Dover.

Jordan, P. W. (2000). *Designing pleasurable products: An introduction to the new human factors*. Boca Raton, FL: CRC Press.

Jung, C. G. (1999). *Essential Jung: Selected writings*. Princeton, NJ: Princeton University Press.

Juutinen, S., & Saariluoma, P. (2007). *Usability and emotional obstacles in adopting e-learning: A case study*. Paper presented at the IRMA International Conference, Vancouver, Canada.

Kahnemann, D. (1973). *Attention and effort*. Englewood Cliffs, NJ: Prentice-Hall.

Kahnemann, D. (2011). *Thinking, fast and slow*. London: Penguin Books.

Karniol, R., & Ross, M. (1996). The motivational impact of temporal focus: Thinking about the future and the past. *Annual Review of Psychology, 47*, 593–620.

Kelley, C. L., & Charness, N. (1995). Issues in training older adults to use computers. *Behaviour and Information Technology, 14*, 107–120.

Kim, C. K., Han, D., & Park, S.-B. (2001). The effect of brand personality and brand identification on brand loyalty: Applying the theory of social identification. *Japanese Psychological Research, 43*, 195–206.

Kuniavsky, M. (2003). *Observing the user experience: A practitioner's guide to user research*. San Mateo, CA: Morgan Kaufmann.

Laming, D. (2008). *Understanding human motivation: What makes people tick.* Cornwall: Wiley-Blackwell.

Law, E., Roto, V., Vermeeren, A. P., Kort, J., & Hassenzahl, M. (2008). Towards a shared definition of user experience. In *CHI'08 Extended Abstracts on Human Factors in Computing Systems* (pp. 2395–2398).

Lazarus, R. S. (1991). Progress on a cognitive-motivational-relational theory of emotion. *American Psychologist, 46,* 819–834.

Lazarus, R. S., & Lazarus, B. N. (1994). *Passion and reason: Making sense of our emotions.* Oxford: Oxford University Press.

LeDoux, J. (1998). *The emotional brain: The mysterious underpinnings of emotional life.* New York: Simon and Schuster.

Leikas, J., Ylikauppila, M., Jokinen, J., Rousi, R., & Saariluoma, P. (2011). Understanding social media acceptance and use in the context of technology generations and life-based design. In C. Billenness, et al. (Eds.), *Information sciences and E-society. Proceedings of INFuture2011* (pp. 253–62). Zagreb: University of Zagreb.

Luria, A. (1973). *Working brain.* Harmondsworth: Penguin.

MacLean, P. (1990). *Triune brain in evolution.* New York: Plenum Press.

Marcel, A. J. (1983). Conscious and unconscious perception: Experiments on visual masking and word recognition. *Cognitive Psychology, 15,* 197–237.

Markman, A. (1999). *Knowledge representation.* Mahwah, NJ: Lawrence Erlbaum.

Maslow, A. H. (1954). *Motivation and personality.* Oxford: Harpers & Row.

Matsumoto, D. (2000). *Culture and psychology.* Stamford, CT: Wadsworth.

May, R. (2009). *Man's search for himself.* New York: WW Norton.

Mayring, P. (2000). Freude und Glueck [Joy and happiness]. In J. H. Otto, H. A. Euler, & H. Mandl (Eds.), *Emotionspsychologie.* Weinheim: Beltz.

McCrae, R. R., & Costa, P. T. (1987). Validation of the five-factor model of personality across instruments and observers. *Journal of Personality and Social Psychology, 52,* 81–90.

Mees, U., & Rohde-Höft, C. (2000). Liebe, verliebtsein und zuneigung [Love, loving and affection]. In J. Otto, H. Euler, & H. Mandl (Eds.), *Emotionspsychologe.* Weinheim: Beltz.

Mischel, W. (1973). Toward a cognitive social learning reconceptualization of personality. *Psychological Review, 80,* 252–283.

Mischel, W. (1993). Introduction to Personality. Orlando, FL: Harcourt Brace College Publishers.

Monk, A., Hassenzahl, M., Blythe, M., & Reed, D. (2002). Funology: Designing enjoyment. CHI'02 Extended Abstracts on Human Factors in Computing Systems, 924–925.

Moody, G. (2002). *Rebel code: Linux and the open source revolution*. New York: Basic Books.

Moors, A., Ellsworth, P. C., Scherer, K. R., & Frijda, N. H. (2013). Appraisal theories of emotion: State of the art and future development. *Emotion Review, 5*, 119–124.

Myhill, C. (2003). Get your product used in anger! (Before assuming you understand its requirements). *Interactions, 10*, 12–17.

Nagamashi, M. (2011). Kansei/affective engineering and history of Kansei/affective engineering in the world. In M. Nagamashi (Ed.), *Kansei/affective engineering* (pp. 1–30). Boca Raton, FL: CRC Press.

Neisser, U. (1967). *Cognitive psychology*. New York: Appleton-Century-Crofts.

Newell, A., & Simon, H. A. (1972). *Human problem solving*. Engelwood Cliffs, NJ: Prentice-Hall.

Niedenthal, P. M., Krauth-Gruber, S., & Ric, F. (2006). *Psychology of emotion: Interpersonal, experiential, and cognitive approaches*. New York: Psychology Press.

Nonaka, I., & Takeuchi, H. (1994). *The knowledge creating company*. Oxford: Oxford University Press.

Norman, D. A. (1999). Affordance, conventions, and design. *Interactions, 6*, 38–43.

Norman, D. (2004). *Emotional design: Why we love (or hate) everyday things*. New York: Basic Books.

Oatley, K., Keltner, D., & Jenkins, J. M. (2006). *Understanding emotions*. Malden, MA: Blackwell.

Orlikowski, W. J. (2000). Using technology and constituting structures: A practical lens for studying technology in organizations. *Organization Science, 11*, 404–428.

Ortony, A., Clore, G. L., & Collins, A. (1990). *The cognitive structure of emotions*. Cambridge: Cambridge University Press.

Osborne, J., Simon, S., & Collins, S. (2003). Attitudes towards science: A review of the literature and its implications. *International Journal of Science Education, 25*, 1049–1079.

Panksepp, J. (1998). *Affective neuroscience: The foundations of human and animal emotions*. Oxford: Oxford University Press.

Parasuraman, R., & Rizzo, M. (2006). *Neuroergonomics: The brain at work.* Oxford: Oxford University Press.

Pervin, L. A., & Oliver, P. (1997). *Handbook of personality: Theory and research.* New York: Wiley.

Porges, S. W. (2003). Social engagement and attachment. *Annals of the New York Academy of Sciences, 1008,* 31–47.

Power, M., & Dalgleish, T. (1997). *Cognition and emotion: From order to disorder.* Hove: Psychology Press.

Reinders, M. J., Dabholkar, P. A., & Frambach, R. T. (2008). Consequences of forcing consumers to use technology-based self-service. *Journal of Service Research, 11,* 107–123.

Rogers, W. A., & Fisk, A. D. (2000). Human factors, applied cognition, and aging. In F. I. M. Craik & T. A. Salthouse (Eds.), *The handbook of aging and cognition* (pp. 559–591). Mahwah, NJ: Lawrence Erlbaum Associates.

Rolls, E. T. (2000). Precis of the brain and emotion. *Behavioral and Brain Sciences, 23,* 177–191.

Runes, D. (1983). *Dictionary of philosophy.* New York: Philosophical Library.

Ryan, R. M., & Deci, E. L. (2000). Self-determination theory and the facilitation of intrinsic motivation, social development, and well-being. *American Psychologist, 55,* 68–78.

Saariluoma, P. (1997). *Foundational analysis: Presuppositions in experimental psychology.* London: Routledge.

Saariluoma, P., & Jokinen, J. P. (2014). Emotional dimensions of user experience: A user psychological analysis. *International Journal of Human-Computer Interaction, 30,* 303–320.

Salehi, B., Cordero, M. I., & Sandi, C. (2010). Learning under stress: The inverted-U-shape function revisited. *Learning and Memory, 17,* 522–530.

Sanders, D. J., & Wills, M. F. (2005). *Cognitive therapy: An introduction.* London: Sage.

Santos-Pinto, L., & Sobel, J. (2005). A model of positive self-image in subjective assessments. *American Economic Review, 95,* 1386–1402.

Scherer, K. R. (2005). What are emotions? And how can they be measured? *Social Science Information, 44,* 695–729.

Schmitt, A., & Mees, U. (2000). Trauer [Sorrow]. In J. H. Otto, H. A. Euler, & H. Mandl (Eds.), *Emotionspsychologie.* Weinheim: Beltz.

Seligman, M. E. P., & Csikszentmihalyi, M. (2000). Positive psychology—An introduction. *American Psychologist, 55,* 5–14.

Shelton, B. E., Turns, J., & Wagner, T. S. (2002). Technology adoption as process: A case of integrating an information-intensive website into a patient education helpline. *Behaviour and Information Technology, 21*, 209–222.

Spinoza, B. (1675/1955). *Ethics*. New York: Dover.

Stenros, A. (2005). *Design revolution*. Jyväskylä: Gummerus.

Switzer, F. S., III, & Sniezek, J. A. (1991). Judgement processes in motivation: Anchoring and adjustment effects on judgment and behavior. *Organizational Behavior and Human Decision Processes, 49*, 208–229.

Tapscott, D., & Williams, A. D. (2008). *Wikinomics: How mass collaboration changes everything*. Harmondsworth: Penguin.

Teo, T. S. (2002). Attitudes toward online shopping and the internet. *Behaviour and Information Technology, 21*, 259–271.

Tibballs, G. (1999). *Business blunders*. London: Robinson.

Toates, F. M. (1986). *Motivational systems*. Cambridge: Cambridge University Press.

Tversky, A., & Kahnemann, D. (1974). Judgement under uncertainty: Heuristics and biases. *Science, 185*, 1124–1131.

Venkatesh, V. (2000). Determinants of perceived ease of use: Integrating control, intrinsic motivation, and emotion into the technology acceptance model. *Information Systems Research, 11*, 342–365.

Weiner, B. (1985). An attributional theory of achievement motivation and emotion. *Psychological Review, 92*, 548–573.

Wells, A., & Mathews, G. (1994). *Attention and emotion*. Hove: Erlbaum.

Wiggins, J. S. (1996). *The five-factor model of personality: Theoretical perspectives*. New York: Guilford Press.

Williams, J. M. G., Watts, F. N., McLeod, C., & Mathews, A. (1997). *Cognitive psychology and emotional disorders*. New York: Wiley.

Yi, M. Y., & Hwang, Y. (2003). Predicting the use of web-based information systems: Self-efficacy, enjoyment, learning goal orientation, and the technology acceptance model. *International Journal of Human-Computer Studies, 59*, 431–449.

Zimbardo, P. G., & Leippe, M. R. (1991). *The psychology of attitude change and social influence*. New York: McGraw-Hill.

6

Life-Based Design

Technical artefacts should exist to bring added value and quality to people's lives. Human-Technology Interaction(HTI) design should, therefore, be considered in a much broader context than merely the usage of technology. It should be based on an understanding of people's lives and well-grounded design methods and tools, which can investigate life and apply this knowledge to the design work. The conceptual model of life-based design (LBD) is based on segregating unified systems of actions called forms of life. Investigating the structure of actions and related facts relevant to particular forms of life, in addition to the values that people follow, is the core tool of LBD. The knowledge produced constitutes a template for human requirements, which serves as a basis for design ideas and technological solutions.

The motive for using any technology must be sought by investigating the role and influence of particular technical artefacts in people's lives. Regardless of whether the technology is produced for daily activities, work, education, or leisure, the justification for technology is always—directly or indirectly—in its capability to improve the quality of life. Technologies are intended to help people realize their action goals in life, and thus make life easier or richer. Depending on the case, this means that

© The Editor(s) (if applicable) and The Author(s) 2016
P. Saariluoma et al., *Designing for Life*,
DOI 10.1057/978-1-137-53047-9_6

with the help of technology, their situation in life is either improved or maintained, which would not be possible without the given technology.

In the early stages of design, it is common to envision and design what the technology in question should do and how it is intended to carry out its task in practice. The early design phases should be dominated by the question of *what technology is designed for*, which is implicit in any design process. It is not possible to design a technology without at least a fair idea of its future use, although some modern technologies such as social media let users choose the ultimate contents of the technology (Pahl et al. 2007; Ulrich and Eppinger 2011). This 'leave it up to the users to decide' mentality is not, however, suitable in most cases of technology design. The target of the design should be fully understood in order to design purposeful technology.

The present directions of early-stage (or front-end) design can be roughly divided into three approaches based on their focal design concepts. These need-based, action-based, and user-based approaches are applied to explicate the purpose and usage of the designed technology. *Technology need-based thinking* is common in traditional engineering (Leikas 2009; Pahl et al. 2007; Ulrich and Eppinger 2011). The idea is that technology satisfies a human need, for example, to drill holes in wood. The task of the design is to find the right combination of functions and structures for the artefact to satisfy the user's need (Pahl et al. 2007; Ulrich and Eppinger 2011). Thus, solving interaction problems is guided by the knowledge of user need, such as how the tool should be used.

The increasing complexity of technology has challenged designers and design theorists to generate methods that would describe in detail the immediate use of a technical artefact. Good examples of these *action-based approaches* are activity theory, scenario-based design, contextual design, and related ways of thinking, which begin by investigating what people do in their work environment and with technology (Beyer and Holtzblatt 1997; Carroll 1997; Kuutti 1996). For example, when designing an information system for a company, action-based approaches analyse the organizational activities and, based on the results, create technical solutions to support company personnel in their work.

The rapid development of everyday computing has raised interest in the properties of users and, accordingly, *user-based* or *user-centred* approaches. In such design approaches as *goal-directed design*, design with *personas* (Cooper et al. 2007), and *human-driven design* (Brand and Schwittay 2006), users as individuals are considered the cornerstone of the design process. Designers generate descriptions of potential end users and their use contexts to obtain an accurate picture of how the technology should be used and what kinds of design requirements or targets for the particular technical artefact or service should be created.

The curt introduction of the design approaches above is meant to highlight that the differences between them are not imperative, and that instead the approaches overlap to some extent. What is significant is that none of these approaches has reached a sufficient level of generality to assimilate others. Each of them is focused on a restricted area of the interaction phenomenon, and they have many elements in common but also differences. While they do not offer a holistic view of the design, together they conceptually cover most of the known design issues.

It is possible to examine the questions of 'what for' from a more focused and wider perspective than that of the immediate usage situation. Instead of focusing on the use of the technical artefact, it is reasonable to focus from the very beginning of the design process on designing *life* itself, that is, how people can improve the quality of their life. Instead of only paying attention to the technical artefact, one should also focus on analysing, investigating, and designing how people live. Different artefacts are important in the sense that they enable people to carry out tasks. However, the first goal of technology design should be to understand what these tasks should first consider how life itself can be changed. The second goal, then, is to create tools for that change. Life provides a solid concept for interaction design processes that arise from the question 'what for'?

Life in this context refers to work life, daily activities, as well as leisure time, all of which should be considered when exploring what people need in life in terms of and roles for future technologies. This chapter develops the foundations for this design approach—LBD—that unites the need-, individual-, and action-based approaches and thus provides general foundations for front-end design (Leikas 2009; Saariluoma and Leikas 2010).

When life has been accepted and adopted as the basis for technology design thinking, it is necessary to move on to construct the basic structure of related design thinking using basic concepts, procedures, and questions. The goal is to outline practical contents for designing for life. Critical questions include how should life be conceptualized, what are its design-relevant aspects, what are the scientific foundations for designing life and how can the knowledge of these aspects be used to outline a general HTI design framework.

LBD and Forms of Life

All scientific approaches must define their basic concepts and research topics. Yet it is difficult to find limiting definitions to describe life in a unified, unambiguous, and satisfactory manner in the context of technology. Even the very concept of life itself is difficult to define in this context, as it is a phenomenon that all people intuitively know exists, and which they experience every second.

Biology is said to be the science of life, and it has been assumed that it defines what life is comprised of (Mayr 1998). However, human life is only partly understandable in biological terms. Life, as it should be understood in the context of technical artefacts, also comprises the different practicalities of everyday experiences as well as socio-cultural and mental aspects.

The concept of life is a *conceptual postulate* such as a 'system' or 'function' (Saariluoma 1997) that can be applied in many ways depending on the context. Like most conceptual postulates, 'life' is a family resemblance concept (Wittgenstein 1953), which can be used with many different, loosely connected meanings in different contexts (Wittgenstein 1953). There is certainly a significant difference between, for example, such notions as 'the life of unicellular creatures' and 'social life'. Technology is not central for unicellular creatures; there are many complex technological systems that support people's social lives. Thus, when designing an artefact, it is difficult to think of life as a whole. Instead, it must be broken down into definable parts to find a proper concept for the purposes of the design target. Design requires a concept that can be used to separate any concrete context of human life from all other possible contexts of life. *Form of life* is a concept that separates the design contexts (Leikas 2009; Leikas and Saariluoma 2008; Wittgenstein 1953).[1]

The main function of the concept 'form of life' is to define and separate the section of life for which the new technology is designed, and to help understand the contents of this particular area of life. Form of life is, thus, a general concept used to abstract *any system of actions in human life* under scrutiny and design. This concept has been used as a part of sociological discourse (Giddens 1984, p. 74, 2000), but should not only be studied from a sociological perspective. In addition to social elements, human forms of life are determined and shaped by many biological and psychological factors. Even though these three form the basic human life sciences, they all have internal and combined fields of research that are relevant for the analyses and theoretical discussions in LBD.

Form of life is thus a general conceptual abstraction, but it enables designers to define the topic and organize their research. Such conceptual abstractions are a common procedure in all sciences. For example, a chemist may be interested in the properties of molecules. Before it is possible to proceed with the research, it is necessary to define whether the research topic concerns hydrogen, sodium chloride, or something else. Similarly, before a form of life can be studied, it first has to be distinct. Whereas chemists can be interested in pentane molecules rather than other molecules, a life-based designer may be targeting solutions at, for example, recreational travellers as opposed to businessmen. Ultimately, researchers in both fields have to decide and express their topics in a similar manner.

Form of life is a highly flexible concept, yet at the same time it is also exact. Using it, one can refer equally well to a medical doctor's ways of working as to the hunting habits of a little-known tribe. People's lives, to a great extent, are characterized by the different kinds of regularities they

---

[1] The original German term for 'form of life' is 'Lebensform' (e.g., Wittgenstein 1953: §19). The concept *form of life* originates from Wittgenstein's (1953, 1964) late philosophy. By this term, Wittgenstein, one of the most important philosophers of the last century, refers to any circle or context of linguistic actions. In his original proposal, form of life was a theoretical concept and conceptual abstraction for analysing human linguistic behaviour and use of language. It is possible to extend the use of this concept to analyse, for example, human social life (Giddens 1990) and any other aspects of human life (Leikas 2009; Saariluoma and Leikas 2010).

follow. For some people, these may include hunting and preparing the catch, for others driving to work every weekday or having dinner with friends once a month.

People can participate in an unlimited number of possible forms of life. A form of life can be a hobby or activity, profession, family status, or a situation. Students, weight-watchers, voluntary workers, teenagers, golfers, criminals, policemen, bankers, designers, Finns, Spaniards, slow-food lovers, grandmothers, and alcoholics all have different forms of life, and they may find new forms of life or develop old ones. Thus, everyone participates in several different forms of life, and their regularities give meaning to people's actions and aims.

It is important to note that form of life is not identical to the concept of culture. Cultures explain many differences in forms of life, from, for example, designing clothes to participating in political or religious ceremonies, but not all characteristics of forms of life are culturally motivated. For example, it is not a cultural but a biological phenomenon that people age and start to experience age-related decline in physical functional abilities. This change in physical abilities, however, may change people's daily life and consequently create new forms of life, such as utilizing rehab services or moving to a senior home, which entail culture-dependent triggers and modifiers (Leikas 2009). To give another example, a compass in a mobile phone that indicates the direction of Mecca is based on a cultural form of life. Similarly, toys for children playing in a sandbox are based on a special form of life. Thus, everything that people do takes place in a definable form of life. For this reason, the first step in LBD is analysing the form of life (Leikas 2009; Saariluoma and Leikas 2010).

Participating in a form of life is not necessarily a voluntary choice. Instead of choosing it, an individual may be 'thrown' into it. For example, a woman born in Scandinavia might have a different conception of family roles than a woman from Africa. The official and public, as well as private, individual, and tacit regularities in a human life's actions form logical wholes that constitute forms of life.

Understanding a particular form of life helps develop technical solutions for people participating in it. For example, designing for young and sporty people is very different from designing for retired but active seniors. Although these groups have the common denominator of 'active life', their forms of life differ in many ways. A form of life is, thus, a tool for exposing relevant differences between different life settings. It provides a precise yet elastic enough concept for an LBD process to define its target.

Structuring a Form of Life

By defining and analysing forms of life, it is possible to understand what kinds of entities everyday life contains. People *participate* in forms of life by sharing different rules and regularities (Wittgenstein 1953). When ice hockey fans, for example, go to the ice stadium, or get together with other fans in a pub, watch a match on TV or read sports news, they participate in the hockey fans' form of life. They carry out certain actions that are typical of this particular form of life.

The organized systems of regular actions in a form of life—in Wittgenstein's (1953) terms *rule-following actions*—are connected. Summer cottagers perform typical actions when participating in this form of life: they usually refuel and pack the car, stop at a grocery shop or supermarket, drive to the cottage, unload the car at the destination, look around that everything is all right, heat up the cottage, sweep the courtyard, make things comfortable, and relax. This form of life involves quite different types of rule-following actions from that of a cultural traveller, for example. These facts are significant when specifying the rule-following actions from the point of view of a specific group of people.

The notion of rule-following actions provides a focus when analysing forms of life (Leikas 2009; Saariluoma and Leikas 2010), and the first step in investigating any form of life is to consider what the main rule-following actions are (Saariluoma and Leikas 2010). The school child's form of life can be defined by investigating what kinds of regular actions are undertaken during a school day. Children wake up, get washed, have breakfast, go to school, take part in schoolwork, meet classmates during breaks, return home, and finally do their homework. These regularities are rules that children follow by acting in a regular manner when they participate in the school child's form of life (Latvala 2006).

With the help of analysing forms of life and their structure of rule-following actions, it is possible to innovate technology that would helps people in their actions and practices. It can be said that rule-following actions have their meaning in a form of life, and that the rule-following actions and their goals give a meaning to technological ideas, and consequently, to designing technology.

Form of life considers the activities of a *group of people* instead of individuals. This premise makes it possible to found the development of technology on this notion. Naturally, from an individual perspective, people can have a certain way of life, but examining a form of life always binds the point of view to a larger group of people. In this respect, forms of life can be examined, for example, through certain generations with a specific nationality and background.

When carrying out rule-following actions, people do not necessarily follow the rules any more consciously than when they encode letters when reading a text. The rules may equally well be subconscious patterns of behaviour, such as rules of conduct and etiquette that people follow when visiting each other. Many rules are explicit, such as tacit rules at the work place—and some may even be formally regulated or juridically determined, such as traffic codes—but most often they are just ways of acting in life. It may also be that people may take different actions to follow the same rule. One of Wittgenstein's (1953) key remarks in using the notion of rule-following actions is that these actions are regular but not mechanistic. People can violate, reject, or even neglect rule-following actions and still participate in a form of life. Soccer fans do not need to watch every match their favourite team plays, children do not always have homework, and one need not always visit a grocery shop on the way to the summer cottage. It is also possible to reach these goals in different ways. So, although they are regularly followed, rules do not have to be absolute.

Mere discrimination of rule-following actions of a form of life does not provide sufficient information. For early-phase design purposes, it is essential to understand the internal structure of rule-following actions. One has to see how actions are integrated in order to define the relationships between them, extract their similarities, and explicate the logic behind them (Leikas 2009; Saariluoma and Leikas 2010).

The Structure of Rule-Following Actions

The next step in investigating a form of life during a design process is to consider what kinds of actions people carry out and what kinds of integrated systems of rule-following actions or deeds are included in the form of life. The first step in doing so is to examine the structural elements of all human actions (Parsons 1968) (Fig. 6.1).

One of the major Finnish historical novels of the last century, *Under the North Star*, begins with the sentence 'In the beginning there was the swamp, the hoe—and Jussi' (Linna 1959–1962). Implicitly, one realizes that Jussi has a task: he should drain the swamp and turn it into a field of grain and he has a tool—hoe. Upon reflection, the ontological structure of the actions of the settlers of Jussi's time are similar to that of today's schoolteachers. The latter have a task—to transmit knowledge to pupils (cf. how to drain a swamp), they are agents (like Jussi) and they have tools (pads, chalk, classrooms etc.). Although the contents of tasks, and the tools and skills for accomplishing the task, are different, the structure of actions has remained the same.

Fig. 6.1   Main components of human action

Defining the conceptual structure of human actions when using an artefact starts by defining the goal of the action (for an alternative analysis of actions, see, for example, Parsons 1968). Second, in every action, there is someone who acts—in other words, the action's agent. Third, there has to be an artefact or tool with which the action is realized. Fourth, the action requires a target. Finally, an action cannot be executed without a usage environment, that is, a context, in which it is carried out. This five-component schema of action explicates the conceptual structure of any situation in which a technical artefact is used (Fig. 6.1).

When using technology, actions have *goals*, as their purpose is to alter the existing state of affairs. This alteration may be a change in a physical state. In Jussi's time, as on cattle farms today, it is important to fence cattle. Digging a hole in the ground makes it possible to put up a pole and then build a fence. The new state of affairs—a hole in the ground—is the goal of the actions, and these actions aim to cause a change in the existing physical environment.

It is good here to call readers' attention to the difference between what people *do* and what *happens* to them. If action is goal directed by nature, it is considered as 'doing' rather than merely causal happening. If a human being falls from the top of a building, this can be a consequence of stumbling, which means that the person originally had a different intention. Falling is then considered an action that happened *to* the person. But if the person jumps in order to commit suicide, it is an intentional action carried out in a goal-directed manner and is thus considered *doing*.

An *agent*—that is, the acting self—refers to the human being or beings who are carrying out actions (Searle 2001). The subject of the action can also be called a *user*. Human roles in using technologies can vary. Often people are *active users* and they steer the technical artefact towards the expected goal. They can also be *passive users*, such as travellers on a train or in an aeroplane. Finally, they can be *objects* for using technical artefacts, such as patients at a dentist or citizen records in national information systems. Without an agent, no action would be possible. The different properties of an agent, such as age, skills, profession, gender, and intention can have a significant role in deciding what kind of technology is required and what would be purposeful for the design of a particular artefact (Cooper et al. 2007).

Naturally, a significant conceptual attribute of using any technology is the *technical artefact* itself (Gero 1990), which have different functions, behaviours, and structures (components, links, and operative logic). Technical artefacts serve to achieve the action goal. Their behaviour is predetermined, that is, it is decided what kinds of states they can reach and what kinds of processes they follow to accomplish the expected goal compared to a given initial state. They include attributes that describe their behaviour, such as effectiveness, purposefulness, and ecology.

The fifth conceptual element of using technology is *target*. It constitutes the elements caused by the action that need to be changed, modified, or removed from the environment. The target is dependent on a *context* of action, and a context is a precondition of the actual performance to take place. This component of interaction comprises a large entity of different perspectives in the usage environment. The *physical context* is formed by different kinds of physical phenomena such as illumination, temperature, noise, or wind. A heavy wind, for example, may affect the way an aeroplane behaves, and direct sunlight may affect the visibility of a display. Contexts can also include social, psychological, technical, and task-based elements. A *social context* includes different stakeholders that are indirectly involved in the usage of an artefact, such as work mates in the case of office technology, family members of a patient in hospital, virtual friends in social media, or legislation behind home care services for elderly people.

Information forms its own context, which becomes increasingly important in designing technology. Naturally, even Jussi in the example above needed to understand how to dry the swamp. He had to have the know-how to be able to dig a ditch and to place the stones properly in order to direct the water away from the swamp. The modern world is filled with information contents relevant for carrying out any action. The worldwide web, big data, Internet of things, and social media surround people and form a central part of their action environment. These kinds of information contexts form an additional and independent element of current and future technology.

In context design, all circumstantial effects of the context must be taken into account. Sometimes it can be difficult to even distinguish the difference between the target and a context. For example, technology that is designed to change a temperature in a given environment is actually modifying a context at the same time as it accomplishes the target of the action.

Goal, agent, artefact, target, and context together describe the highest conceptual and ontological schema of any situation in which people use technology. The properties of these concepts introduce task-necessary questions, which must be answered either explicitly or implicitly. They form the framework in which HTI design problems are solved.

Understanding and Explaining Forms of Life: Facts and Values

The logic behind rule-following actions is valuable information, as it reveals why rules make sense, what the goals of actions are and how the actions are integrated. However, rules and regularities in actions are not the only significant factor in understanding a form of life. *Facts* and *values* also help explain why people follow these actions.

People's lives can be surrounded by many factors such as age, gender, marital status, health, standard of living, size of household, religion, nationality, education, and work history. They can all be considered *facts in life* and attributes of a form of life. They explain (and often even determine) the goals of rule-following actions and provide the limits—the possibilities and restrictions—for what people can do in their life.

Facts make it possible to understand people's everyday contexts and the kinds of real needs that arise from these contexts. Along with facts, one has to pay attention to the values that people follow in life in order to create added value for people and improve their quality of life. In this way, the design focuses on opening up possibilities and developing meaningful services for people.

Examples of key questions that can be answered by examining facts and values include: Why does it make sense to have this particular form of life? Why have people developed it? What aim does the form of life serve? What do people gain by participating in it? Why does it have all these regularities? How and why are rule-following actions linked to each other?

Travelling, for example, is quite a different event for a teenager travelling in Europe compared to that of a retired person on a city trip. On a superficial level, these groups of people seem to follow the same system of rules—the rules of travelling—but on a deeper level, the divergent frameworks composed of facts in their life make their travelling events quite distinct from each other.

The first group of facts in life that signifies, for example, the difference between young and older people is the *biological facts of life*, which in the case of a retired person can include a decline in physical functional abilities, such as vision and hearing. This may narrow their opportunities to travel. The second group is *socio-cultural facts*. In the case of older people these may include an increased amount of free time after retirement, and thus the freedom to travel at any time of the year. Another sociological fact is that as people get older, the extent of their social networks—and thus the number of potential travelling companions—usually decreases. The older travellers in Europe usually have more money than younger travellers, and are thus able to organize their trip in a more comfortable manner from their own point of view. In other words, their social conditions are different from those of young travellers. Young travellers often have a minimal backpacking budget and are keen to find new friends, perhaps even new partners. To save money, they are prepared to use sleeping bags, stay in campsites and youth hostels, and eat cheap food. Many older travellers could not, for health reasons, travel in this way.

The third group of facts is the *psychological facts of life*. A psychological fact can be, for example, that older people have more experience in life but often also stronger resistance to change than young people. Another psychological difference between these two groups is skills or expertise (Ericsson 2006). For example, the ICT skills of young and older people may differ. Young people are usually technically more skilled and able to reap more benefits from ICT technology during a trip. They may use technology to find, for example, information about travelling conditions and tourist attractions. They can also take advantage of online maps and other available services, while older people may mainly focus on sending SMS messages and talking on the phone.

The different facts in life of these two groups of people influence their travelling styles and the form that the life of a traveller takes. As can be seen from the above examples, facts in life consist of *biological, psychological,* and *socio-cultural* aspects that influence the constitution of a

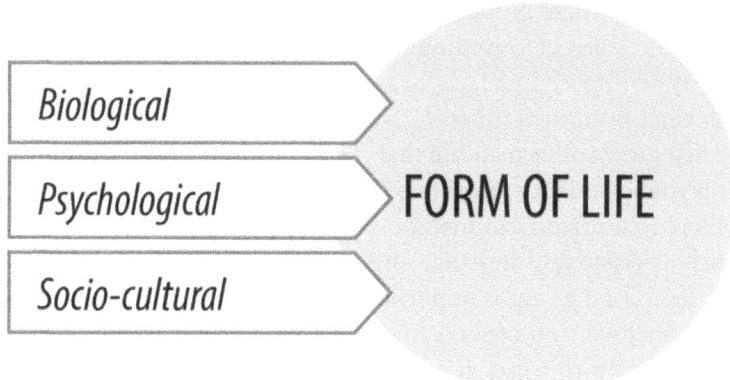

Fig. 6.2   Global explanatory frameworks for the form of life

form of life (Fig. 6.2). As already pointed out, biological facts are essential in explaining the very basic elements of a form of life. Psychological and socio-cultural facts, in turn, arise from the corresponding elements of people's lives. They become visible by modifying the rule-following actions in different forms of life accordingly.

Forms of life are comprised of the biophysical conditions of an individual as well as the customs, habits, rules, and language games of the person in question and many other people. In this sense, people are placed in pre-existing forms of life and have little chance of changing them. Some of the facts in life are more or less 'chosen' by the individual and may be dynamic in nature, whereas others are 'given'—that is, inherited or determined by living conditions—and thus constitute stable factors of the holistic form of life. An example of such a stable factor determined by living conditions is the lives of war-affected children. Thus, form of life is a holistic notion that includes predetermined factors in life as well as a complex variety of different elements in everyday life.

In addition to facts, *values* provide necessary information for analysing and understanding forms of life. They explicate individual or group goals, demands, obligations, and conceptions of beauty and goodness in life, and direct and affect people's behaviour, goal settings, choices, and attitudes. Logically, technologies must be designed to support these actions (things that are valuable to people).

Many elements of form of life culminate, and are reflected in, values. Similarly, values can be shaped by different elements of a form of life. For example, mobile technology has changed the way people look at social relationships. Before mobile phones, social relationships were to a large extent based on certain locations and largely depended on a person's specific environment (Hulme and Peters 2001; Hulme and Truch 2006). With mobile phones, however, it is not immediately obvious to the caller where the individual is calling from. At the same time that the boundaries of environments have become more fluid, privacy conceptions have become more flexible. Along with the phone's value conception of 'anytime, anyplace', people are starting to appreciate and take for granted that they can be available every time the phone rings. Thus, mobile technology has changed the concept of reachability: 'always being available' has become a value for most people. An exception is made by the increasing group of people that is committed to being available because of work demands and that now seeks for possibilities for down shifting in life.

Values are either moral or practical, and they may include personal, social, cultural, religious, philosophical, ethical, and aesthetic dimensions. For example, Catholic people in many countries carry out decorated and carefully organized processions of hundreds of people during the Holy Week at Easter. The religion-based personal and community values of participants of this ancient tradition explain why these people are ready and willing to spend long hours preparing for these festivities over the years, decades, and centuries.

Values as attributes of forms of life are different from facts, in that they represent voluntary choices of the people in question. They influence the selection processes that people carry out between different forms of life, and make people's future plans visible. They also give meaning to actions within a specific form of life. For instance, praying towards Mecca is not meaningful for all people who pray. Instead, knowledge of the direction of Mecca is very valuable to a devout Muslim due to his or her form of life. Based on their values, people give different weightings to different goals, and values can make it understandable why people act as they do.

Understanding values helps identify the kind of 'worth' that technology can bring people (Cockton 2004, 2006). The added value can be attributed to the outcome of the *use of a product*. Moreover, it can be seen as incorporating human values into product development. For example, the value of a product can be analysed in a utilitarian sense: its goodness depends on the happiness or benefit that it brings, and producing a product is considered ethically correct if it brings about the 'maximum benefit' for the largest group of users. Utilitarianism is often felt to be a workable philosophy with respect to new technologies in the sense that the best solution is the one that produces the greatest amount of happiness for the greatest number of people. For example, the development of health care and medical technology can be seen to produce a utilitarian good (maximal value for a maximum number of people).

Through incorporating human values into product design it is possible to enhance the desirability of products and thus increase the value of the product from a commercial and social point of view (Stahl 2006, 2010). Likewise, economic and business value from the point of view of product or service providers should also be emphasized.

Values also call into question what kinds of technologies should be designed for a form of life to improve its value climate in terms of moral standards. A morally virtuous action without expectation of reward is a good in itself. Altruistically, doing something beneficial for others can be considered morally valuable. Moral designers may ask themselves whether their work really promotes 'the good of man'. This is an ethical question that concerns the human consequences of one's work (Bowen 2009). When committing themselves to this aim, designers are faced with a substantial responsibility. Technology will ultimately change the structures of society and the everyday lives of people in a profound way. The kinds of expressions that society will have depend significantly on the design approaches of information society. Ethical design means, first of all, conscious reflection on ethical values and choices with respect to design decisions. Second, it involves reflection on the design process and the choices of design methodologies. Finally, ethical design must consider the issue of what is ethical; that is, what constitutes the good of man. Answering questions of this kind can be supported by knowledge of citizens' forms of life (Leikas 2008, 2009; Stahl 2006).

In ICT research and development, ethical design is grounded on information (or 'computer') ethics, the field of academic study that examines the actual and possible impacts of information and communication technologies on important human values, such as life, health, psychological and physical well-being, happiness, one's abilities, peace, democracy, justice, and opportunities. The overall goal is to advance and defend human values in light of the possible and actual impacts of ICT (Bynum 2010; Stahl 2010).

ICT technology partly constitutes the things to which it is applied. It shapes people's practices, institutions, and discourses in important ways. What health care, public administration, politics, education, science, transport, and logistics will look like 20 years from now will in important ways be determined by the ICT applications that are developed for use in these domains. ICT will also shape the way individuals experience themselves and others.

ICT has introduced many useful things into people's everyday lives, but along with its positive effects its usage may lead (and has led to) many risks, for example, related to information complexity, security, and privacy. For example, ICT applications have provoked ethical discussion in areas such as online medical consultations and home monitoring of older people. Ethical questions related to these issues have concerned, for example, confidentiality, data protection, civil liability for medical malpractice, prevention of harm, informed consent, and patient confidentiality. The development of ICT can also be menaced by faith in the omnipotence of technology. With the use of ICT technology, it is possible to instantly influence other people's worlds from one's home computer.

Different viewpoints on value substantially deepen the analysis of life needed for technology design processes. However, values cannot be separated from the analysis of forms of life, nor should they be separated from rule-following actions. Both facts and values associated with rule-following actions explain the goals to which people aim and the facts that motivate them. Based on this knowledge it is possible to derive goals for technology concepts. It is also possible to give detailed analyses of how the new technology should be realized.

To summarize, consider value construction related to different objects. People are attached to certain possessions because they have constructed values within them, for example, a childhood book or gift from a loved one. This *physical-object attachment* is an important phenomenon. Modern technology has also introduced *digital-object attachment*. For example, the increasing role of online services in connecting people and communicating via different systems and media—in addition to the increasing capacity of digital repositories for personal content and of personal spaces in social networking systems—has created a digital world populated by objects that people can build value with, which become important artefacts of life. Digital objects have thus become cherished possessions to which people feel attached. These new forms of attachment to digital objects could require new technological applications. In order to discover what kinds of applications people would need, the value constructions of people within specific cultural settings need to be studied (Odom et al. 2013).

The next section reflects more systematically on what kinds of design discourses are needed to transform knowledge of actions, facts, and values into concept descriptions of new technology; that is, how to turn this knowledge and understanding of human life into technology requirements.

Technology-Supported Actions

The properties of the concept 'form of life' were analysed above, along with its relationship to the human actions, values, and facts of life that define people's goals. Different forms of life are characterized by different types of rule-following actions typical to a particular form of life. The forms of life of older people include performing daily errands, attending cultural events, visiting friends and relatives, taking care of their own well-being and health, getting medical advice, spending time at a holiday home and travelling, for example (Leikas and Saariluoma 2008). People participate in forms of life by undertaking actions that enable them to follow related rules.

Analysing the form of life is thus the first step in the design process. Based on the analysis of rule-following actions and facts and values in life, it is possible to consider the role of the designed technology in the chosen form of life of the target group. The next question is how this information can be used to design technical artefacts, and help people fully participate in their chosen form of life. The idea of what a technology can be used for provides the focal problem for design, and answering the question 'what for?' thus constitutes a *focal design idea*. The next crucial step is to elaborate the focal design idea and transform the acquired information into technical interactions and concept descriptions. This process relies on the information collected in the analysis of the form of life. Its task is to transform the rule-following actions into *technology-supported actions* (TSAs) (Leikas 2009).

As introduced earlier, rule-following actions, and the forms and reasons behind them, together with information on facts and values, form *design-relevant attributes* in a particular context (Fig. 6.3). After examining the design-relevant attributes, it is clear how some actions can be supported by technology. These target actions constitute TSAs for the particular form of life. They are actions that are realized with the help of technical artefacts. Walking is a rule-following action, and walking with a stick when a person is injured is a simple example of a TSA. Walking sticks are hence technologies that support and modify the original action of walking.

TSAs should consist of elements relevant to the generation of design ideas. These elements are the action and its goal, the agent, the context, and the possible technology. This categorization is characteristic of all human actions, as explained in the previous chapter.

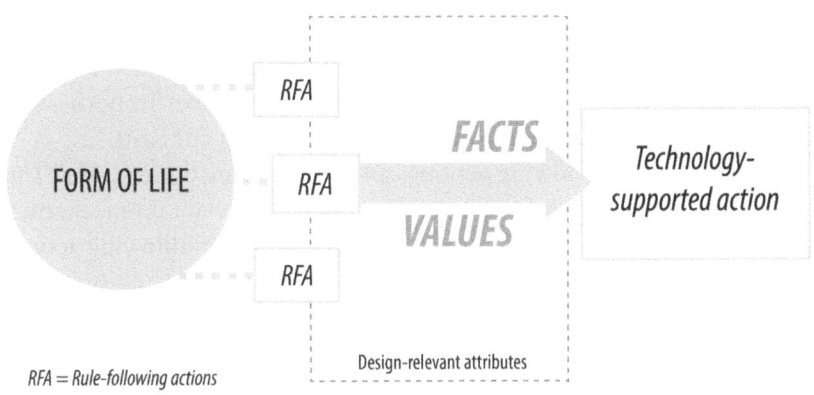

Fig. 6.3 Generation of design ideas from a form of life

For example, before designing a service that could provide information for senior citizens, it is necessary to describe the senior citizens' need for such a service—in other words, their *goals* for using it. This includes factual knowledge of the *agents* (older adults) as users of technology, and their backgrounds, living environments, and rule-following actions. Value attributes also have a central role in this discourse: to extract the goals of the agents and select the best way to reach these goals from the user's point of view. After that, a description of the *artefact* by which the service could be realized is needed. In addition, the *targets of actions* that users are supposed to carry out using the application should be described. Finally, information about the physical, social, and psychological *context* of use is needed to describe the interaction between the artefact, agents, and related stakeholders in TSAs. After elaborating on this information, it would be possible to proceed with a more detailed design concept with the task-based elements of technical contexts.

An example of how rule-following actions can be changed into TSAs is ticket vending machines at railway stations. At many railway stations train passengers are often forced to wait in long queues to buy their tickets. This takes time, and it is irritating for passengers who are in a hurry. From the design point of view, the passenger needs a faster procedure for purchasing a ticket. The change can be made by analysing the passengers' actions and the facts and values involved in buying a train ticket. Based on this information, it should be possible to define technology that could support faster and more convenient purchasing actions. In the case of railways, the results of the step from rule-following actions to TSAs have been ticket vending machines and buying tickets online. In both cases, the traveller can avoid queues and the companies can cut costs.

In this way, rule-following actions can be transformed into TSAs. This is a vital innovation step further in the design. There are no mechanical means to carry out this transformation. Extracting rule-following actions and investigating their facts and values to find problem points gives, however, a solid basis for early phase interaction design.

Many rule-following actions do not need any technological support. However, some technology is commonly used to improve individuals' performance when carrying out a rule-following action. There are thus two kinds of rule-following actions; the ones that can be supported by technology are TSAs.

The description of TSAs takes the design process forward, as it raises and defines sub-problems in the design. For example, if the task is to use an e-commerce service, it is possible to use the explicated TSAs and attribute information to generate sub-problems. These sub-problems would include such questions as how to get people to trust the service, and how to make payment for purchases easy. Such sub-problems emerge automatically when TSAs are analysed. However, the analytical design process can never be endless; at some stage the integration of the results of sub-problems has to begin. This means that each sub-problem has to find its place in the draft concept before it can be stated that all the problems have been solved at the level of a satisfactory product or service concept. The final concept is a full description of an integrated system of TSAs. This description gives the goal of the technology design and includes information that technology designers need to be able to construct the product in a sense-making manner (that is, in a manner that can be easily integrated into the flow of actions in the form of life). The criterion for the goodness of the concept description is whether it fits the TSAs and the form of life it is supposed to support.

Scientific Foundations

In design, one cannot speak about facts unless they are scientifically grounded. LBD relies on human research, and uses the knowledge and methodological bases of such human sciences and research traditions that are directly connected to research on human life. These can be called *human life sciences* (Fig. 6.4). They can have two substantial roles in the design of technical artefacts and systems. First, the concept of human life sciences can be used to explore design issues when searching for clarity with problems and solutions. Second, it can be used to advocate one design solution over another for a particular design purpose. The core question is how these sciences should be organized with respect to design tasks in order to get the best possible support for design solutions. This calls for a holistic and multidisciplinary examination of the design problem at hand. For instance, when people get older, some social decisions, such as retirement, affect older peoples' social lives, and many sociological

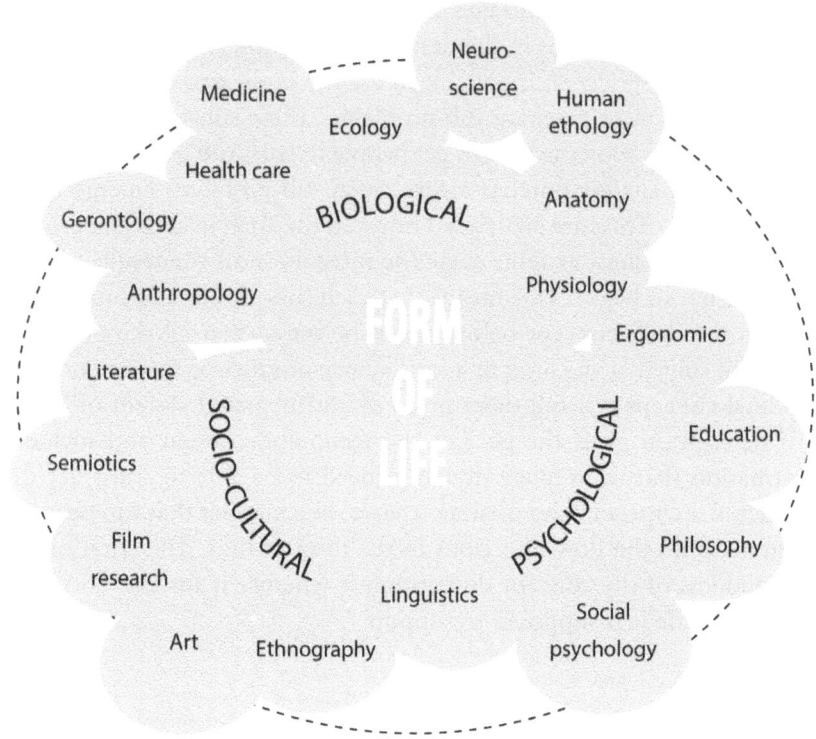

Fig. 6.4   The palette of human life sciences

changes, such as freedom from work and increasing amounts of free time, take place. However, getting old is not merely a social phenomenon. It is also an essential biological phenomenon and has many psychological aspects as well. Therefore, sociological concepts alone cannot give sufficient information about what is relevant in designing modern technologies for older people. It is naturally important to be able to describe the changes in everyday contexts when people age, but it is also necessary to understand the multidisciplinary process of ageing and its different effects. Many of the changes in old age are not caused by social structures, social habits, or social communities, but can be examined, for example, with the help of biological and psychological concepts related to the ageing process.

The relevant research areas of human life sciences are discussed below, starting with *biology and biological life*. These terms refer to all the paradigms of biological and related research that focus on people. Biology introduces numerous perspectives that are important in technology design. People have different biological preconditions for interacting with technologies. Their environment has its biological properties, and human bodies have their biological characteristics, which change over the course of time. Biology has a number of related fields of research, each of which has many potential contact points with modern HTI design. Examples of such research areas can be found in many forms in ecology (e.g., Berkes and Turner 2006; Young 1974), human ethology (Eibl-Eibesfeldt 1989), physiology, anatomy, and neuroscience (Corr 2006), and even in many medicine and health care issues (Coiera 2009).

Without a biological life there would be no life at all. However, biological concepts are not sufficient for understanding all human life and action. Modern brain research has shown that learning, for example, modifies the brain and provides it with 'programs', which cannot be understood merely in terms of biological concepts. Brains are dependent on the properties of learning environments and their contents, and not solely on their own properties (Saariluoma 1999). For example, if a child is born in Germany she will learn German as a native language, and if she is born in South America, her language will most probably be Portuguese or Spanish.

The second dimension of human life sciences is based on the psychology and philosophy of mind, that is, research on people as individuals and the general laws conducting their behaviour. This not only refers to the cognitive aspects of the mind—such as limited capacity or information discrimination—but to all of psychology, including emotional issues such as valence, and personality and social issues such as group behaviour (Saariluoma and Oulasvirta 2010). Again, there are many related disciplines, such as education, ergonomics, and social psychology (e.g., Brown 2000; Karwowski 2006; Nolan 2003).

Sociology, cultural research, and related disciplines form their own cluster of human life sciences (e.g., Argyle 1990; Geertz 1973; Giddens 1984, 1987), which have relevance for design. This cluster includes such research disciplines as social and cultural anthropology, ethnography, gerontology, linguistics, and semiotics, as well as many issues that are usually associated with history, art, literature, and film research.

Sociological concepts often describe the ways people act in society, and thus they fruitfully elaborate the role of human life in design. Social scientific analysis can help understand how and why people participate in different forms of life and share the rules and regularities they involve. This kind of research provides important facts about people and their lives, and it is evident that psychological research, for example, would not provide a similar perspective on life. Instead, knowledge of cultural history or national habits, for example, can be useful in this sense. Many ethical and value questions that are necessary to consider in design also emerge in the context of social and cultural matters (Albrechtslund 2007; Bowen 2009; Bynum 2010; Leikas 2009; Stahl 2010). In addition, there are many interdisciplinary fields of learning that are relevant in HTI design, such as action theory, management, and organizational research (Bannon and Bødker 1991; Kuutti 1996; Nardi 1996). The core challenge is to find a framework that connects the design problems with the respective research.

All the research fields presented above have something to say about human life (Fig. 6.4). They provide methods, concepts, paradigms, models, and theories that may prove relevant in setting, asking, and solving design questions. Indeed, all of these disciplines can play their part in specific types of HTI design processes. In design it is necessary to understand the types of scientific knowledge that are needed to solve different design questions. In order to do this, it is useful to provide an overview of the typical problems concerning the target user group and the knowledge that can be used to design solutions to these problems. After this overview, it is possible to identify the kinds of multidisciplinary concepts that are needed in designing HTI for particular users.

LBD Process

LBD, like all design paradigms, is a methodological process (Leikas et al. 2013). It defines major questions in the field and helps answer them. LBD conceptualizes human life in concepts of human life sciences and connects this knowledge with design thinking.

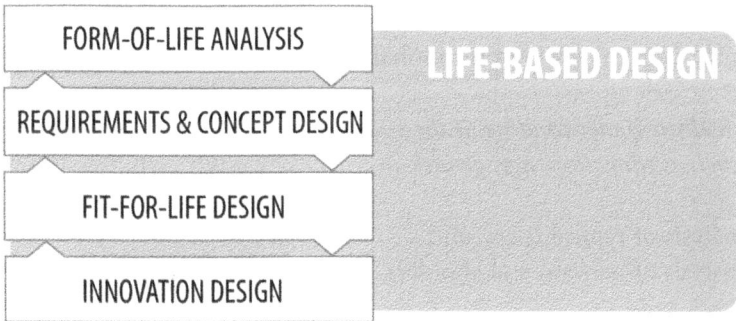

Fig. 6.5   Life-based design model (Leikas et al. 2013)

The LBD model consists of four main phases to guide the designer's thinking during the design process: (1) *form-of-life analysis*, (2) *concept design and design requirements*, (3) *fit-for-life design*, and (4) *innovation design* (Fig. 6.5). Each of the phases can have numerous sub-questions, which are presuppositions for solving the main design questions. The four phases do not have to be sequential; they can be parallel in the iterative design process.

Form-of-life analysis includes defining and analysing the particular form of life the designers are interested in. To produce design-relevant information about the selected form of life, regular actions—the kinds of actions people normally follow—defining the form of life in question have to be extracted. In addition, it is essential to define major explanatory facts and values of the form of life and investigate several issues, as detailed below.

What do people need in their life, and how could life be improved with the help of technology?

* Selection of the form of life in question;
* Analysis of rule following and regular actions;
* Analysis of other relevant characteristic actions; and
* Definition of design goals.

Extraction and examination of the design-relevant human-based problems to define possible problematic side issues:

- Explication of design-relevant problems; and
- Analysis of the existing technology in use.

A realistic understanding of the potential users or actors and their properties, such as education, age, gender, or technology skills:

- Analysis of typical users; and
- Analysis of relevant stakeholders.

Analysis of physical, psychological, and socio-cultural conditions and social relations activated before, after, and when using the technology:

- Analysis of typical contexts; and
- Definition of design theme.

The analytical work should generate the design theme and the *human requirements* of a technical artefact. This information explains the *why's* and *what for's* that should guide the entire design process. Human requirements define how people's life in a specific form of life should be improved. They are based on the methods and results of human life sciences, and form the basis of the next phase in the design by introducing the design theme and the human requirements behind it. However, they do not define the requirements for technological concepts that could be used to address the defined design goals of the specific form of life; these are called design (user) requirements, and they are discussed below.

Concept design and design (user) requirements is the second phase of the LBD model, in which designers define the role of technology in achieving the defined design goals and produce a definition of TSAs in a product or service concept. This means defining what the technology is to be used for (the technical requirements).

An essential part of the concept design phase consists of ideation and outlining what the supposed technological solutions could be like, as well as reflection on (and elaboration of) the selected solution. The outcome of this phase is a definition of TSAs in a product or service concept. It generates prototypes of the relevant new technology. It also explains how, by defining the role of technology in the form of life, this technology can be associated with people's needs, and how technological solutions can be implemented.

A key element of LBD is *co-design* with stakeholders. End users should be seen as experts in their everyday life who can participate in exploring and inventing products and services for their own forms of life. Foundations for co-design will be further elaborated in the final chapter of this book.

The concept design phase ends with technical design and implementation. Here the designers generate mock-ups and prototypes by using the kinds of technologies that should be able to meet the design goals and make it possible for the users to reach their goals in the form of life. In this stage, the issues of user interface, usability, and user experience become focal. These may include, for example, deciding between the use of fixed or mobile technology, as well as special devices for realizing a particular design goal. Congruent with the ISO standards of human-centred design, this phase entails the use of traditional human-centred design processes, such as user interface design and usability and user experience studies as part of LBD (Leikas 2009). These would also include such issues as emotional experience and value of technology for the user (Odom et al. 2013).

To summarize, the main questions of the concept design phase include:

- Defining the role of technology in achieving action goals;
- Solution ideation and reflection;
- Elaborating selected solutions;
- Usability and user experience design;
- User evaluations (usability); and
- Technical design, iteration, and implementation.

The final outcome of this phase is a definition of the technological concept and a description of how people would use it in their specific form of life. Hence, the process explains: (1) how the proposed technology will be part of the user's everyday life and (2) (user) requirement specifications for implementation.

Fit-for-life design is the third phase of LBD. It refers to examining the benefits and meaningfulness users can get from the developed solutions and the impact they have on the quality of life. The outcome of this analysis can lead to improvements and modifications of the product ideas.

This is the most fundamental phase of the LBD framework, as it illustrates *the logic of enhancing the quality of life*. It stresses the importance of returning to (and reflecting on) the human requirements defined in the first phase of the process as well as defining and removing the barriers to implementation. Should there be any problems in fitting the outcome to the form of life, the prototype should be refined according to the requirements. This is the only way to ensure that the outcome truly satisfies the users' needs.

Furthermore, it is essential to bear in mind the ethical and responsible considerations regarding the design concept. Ethical evaluation of the outcome is a natural part of this phase (Bowen 2009; Florini 2010; Stahl 2010). As ethics defines what can be considered the 'good of man', ethical analysis may help explore from whose perspective (and by what kinds of choices) it could be possible to generate an increase of goodness and develop products with a higher value in improving the quality of life.

To summarize, the essential parts of this phase are:

• Illustration of the logic to enhance the quality of life;
• Fit-for-life evaluation; and
• Ethical and responsible design.

Innovation design is the final phase of LBD. It introduces a procedure for exporting the design outcome into general use and incorporating the new technology into human life settings. This process entails activities that are meant to transform the design concepts into products that are used in everyday life. Good design is a prerequisite for ensuring that the future product is really used by people, and therefore innovation diffusion processes are a vital part of LBD.

To make something an innovation, one has to create usage cultures for design outcomes. To ensure that people can in fact use the new technology to realize their goals calls for well-designed guidance and training possibilities. It may also require sophisticated educational programmes. A proper usage culture requires skilled users, and indicates that users can find a role for the given technology in their lives—and know how it could improve the quality of their lives.

Innovation design consists of such sub-questions as definitions of the infrastructure (social and technical), a marketing plan, and service and auxiliary activity. These are methods of communicating the benefits of using the technology. The outcome of the innovation design process is a product that has found its users, or at least a clear plan of how this goal can be reached. This final phase of LBD defines a procedure for exporting the outcome into general use.

The essential parts of innovation design include:

* Creation of a usage culture;
* Definition of the infrastructure;
* Marketing plan;
* Definition of service and auxiliary activity as well as corporate responsibility; and
* Illustration of the technology life cycle.

The process introduced above presents an overview of the significant issues in any technology design process. They emphasize that technology design is not only about techniques and artefacts; it is essentially about designing to improve the quality of human life. The sub-questions raised are not the only ones to emerge during a practical design process. In addition, there are case-specific questions in each design project.

Technology Modifies Forms of Life

As discussed above, technology gains its meaning in the contexts of everyday life. People follow the rules of their adopted forms of life to reach their goals, and the technologies they use are meaningful to the extent that they do so. Technologies enable people to carry out tasks easier, and they often make it possible to do things that would otherwise be impossible. History shows that technology has influenced forms of life in many different ways, both positively and negatively (Bernal 1969).

Forms of life are systems of integrated and organized actions, called rule-following actions. Similar concepts can be found in the literature; one could use such social scientific terms as 'practices', 'traditions', or 'habits' as well. The term action, expressing the intentionality of mind, is more meaningful in the context of design, as it does not imply a longer commitment to this mode of behaviour. Habits and practices presuppose a regular role in individuals' lives, but it is possible to participate in rule-following actions only once. The main issue is to follow a regular action pattern.

To be able to target technology design effectively, it is necessary to have knowledge about forms of life and how they could be improved. This includes understanding why a particular form of life exists and why certain actions of people are relevant. This necessitates knowledge of facts and values associated with the form of life. It has long been known that technological developments change human social life. Technology can also contribute to the extinction of old (and the emergence of new) forms of life. The domestication of animals and the invention of the plough, steam engine, telegraph, radio, nuclear technology, computers, and mobile technologies have all essentially changed the way people live their everyday lives (Bernal 1969; Fichman 2000).

Technologies have made things possible for people, but people themselves have found respective social forms of living. The shift from an agricultural to a modern industrial society did not happen smoothly and easily. It had many social consequences, which changed people's forms of profoundly. In the same sense, applied artificial intelligence and decision-making systems, information networks, and mobile technology are altering what people do, and how and when they do it (Augusto and McCullagh 2007). For example, work life has changed profoundly with the development of information networks, and especially wireless and mobile technology. Company intranets enable distributed and global operations, and mobile devices and the supporting infrastructure allow for new forms of working. E-work is a normal procedure at many work places, including mobile work, remote work, and home-based work (Harpaz 2002). Work tasks are carried out increasingly in global network-like communities and teams. Work can be taken with the employee anywhere, and that person can be connected to his work from almost any place. Colleagues, customers, partners, producers, and competitors can be found in all corners of the global world. Accordingly, along with global networks operating around the clock, the concept of working time has changed.

In addition to shaping behaviour and attitudes, technology affects the way people experience themselves and others. On the basis of their forms of life and experiences, users also modify and alter how they use technology, and as a result may create totally new meanings for different technologies. For example, mobile technology has changed people's attitudes towards devices. A smartphone can be a highly personal device, perhaps more personal than any other technical artefact. The owner of the phone has a personal number, a personal physical device with personal contents, and personalized tone rings. Hulme and Peters (2001) even argue that users consider their mobile phone an extension of the self. Thus, the loss of a mobile phone is felt not just as the loss of a device, but it is also sensed on the level of one's physical self. Indeed, when leaving home without a mobile phone, many people feel that something is missing. So, as users of technology people have come a long way from being considered extensions of machines. The artefact now, at least in the case of the smartphone, is considered an extension of the human.

Although technology has a largely beneficial effect on forms of life, ICT technology does have some downsides. First, the development of ICT has led to a divided society, creating a *digital divide* between technological haves and have-nots. The socio-cultural reasons for not using new technological products or services include ignorance of the services offered; inability to use the services because of lack of knowledge, education, and training; and reluctance to acquire and try new technologies. For example, many commonly used types of interactions are beyond the competence (and thus out of reach) of some groups of people. For some people, the operation logics of new technological solutions are incomprehensible. Moreover, people may not be able to access (or are uninformed about) the services offered through technology. In some cases they are also reluctant to invest their time and effort in trying to learn to use new solutions, especially if they have had bad experiences in using other products or services.

To gain from the benefits of technology and lessen the risks, good design practices are needed. LBD takes the concepts, methods, theories, and empirical knowledge provided by human life sciences as its starting point to serve as grounds for design and research actions as well as tools of thought. It has created a shift in the focus of design thinking: the question is no longer only about technical artefacts and natural sciences; it is also about how people live their lives and how they can improve them. Technologies—when understood as combinations of human action and the technical artefacts or systems in use—are always designed for the well-being of people, and are used to improve the quality of everyday life (Leikas 2009; Saariluoma and Leikas 2010). Therefore, they only make sense in human and social contexts. Because of this human dimension in technology design, it is essential to rethink the nature of design, and early design in particular, of HTI. This shift in focus emphasizes designing life and human life sciences as the basis of technology design. The focus of design and innovation is therefore not on artefacts, but on their capacity to make people's lives richer.

The sub-problems and solutions of designing 'the good life' are case dependent, and can be either conflicting or complementary. *Conflicting sub-problems* cannot be used in the final solution because they would require different functional solutions in similar situations. For example, the designer of a web page must choose a single location for the input fields for credit card payment on the page.

The LBD process therefore requires choosing between plausible alternative solutions. This can be done on the grounds of information about the design attributes. For example, one of the possible solutions may be less usable or ethical than others, and must thus be changed or eliminated.

A challenge—but also a significant strength of—the practicality of the LBD paradigm is the multidisciplinary nature of its applicability. The design should be carried out in a multidisciplinary design team, in which different designers with different expertise work together (and co-design with) different stakeholders. These designers with different backgrounds, perceptions, know-how, and skills should be eager to adopt different design policies and provide a better information society. The goal of LBD is to provide tools for this kind of thinking.

References

Albrechtslund, A. (2007). Ethics and technology design. *Ethics and Information Technology*, 9, 63–72.

Argyle, M. (1990). *The psychology of interpersonal behaviour*. Harmondsworth: Penguin books.

Augusto, J. C., & McCullagh, P. (2007). Ambient intelligence: Concepts and applications. *Computer Science and Information Systems/ComSIS, 4*, 1–26.

Bannon, L. J., & Bødker, S. (1991). Beyond the interface: Encountering artifacts in use. In J. M. Carroll (Ed.), *Designing interaction: Psychology at the human-computer interface* (pp. 227–253). New York: Cambridge University Press.

Berkes, F., & Turner, N. J. (2006). Knowledge, learning and the evolution of conservation practice for social-ecological system resilience. *Human Ecology, 34*, 479–494.

Bernal, J. D. (1969). *Science in history*. Harmondsworth: Penguin.

Beyer, H., & Holtzblatt, K. (1997). *Contextual design: Defining customer-centered system*. Amsterdam: Elsevier.

Bowen, W. R. (2009). *Engineering ethics: Outline of an aspirational approach*. London: Springer.

Brand, P., & Schwittay, A. (2006, May). The missing piece: Human-driven design and research in ICT and development. In *Information and Communication Technologies and Development, 2006. ICTD'06* (pp. 2–10). New York: IEEE.

Brown, R. (2000). *Group processes*. Oxford: Basil Blackwell.

Bynum, T. (2010). The historical roots of information and computer ethics. In L. Floridi (Ed.), *The Cambridge handbook of information and computer ethics* (pp. 20–38). Cambridge: Cambridge University Press.

Carroll, J. M. (1997). Human computer interaction: Psychology as science of design. *Annual Review of Psychology, 48*, 61–83.

Cockton, G. (2004). Value-centred HCI. In *Proceedings of the Third Nordic Conference on Human-Computer Interaction* (pp. 149–160).

Cockton, G. (2006). Designing worth is worth designing. In *Proceedings of the 4th Nordic Conference on Human-Computer Interaction: Changing Roles* (pp. 165–174).

Coiera, E. (2009). Building a national health IT system from the middle out. *Journal of the American Medical Informatics Association, 16*, 271–273.

Cooper, A., Reimann, R., & Cronin, D. (2007). *About Face 3: The essentials of interaction design*. Indianapolis, IN: Wiley.

Corr, P. (2006). *Understanding biological psychology*. Malden, MA: Blackwell.

Eibl-Eibesfeldt, I. (1989). *Human ethology*. New York: de Gruyter.

Ericsson, K. A. (2006). The influence of experience and deliberate practice on the development of superior expert performance. In K. A. Ericsson, N. Charness, P. J. Feltovich, & R. Hoffman (Eds.), *The Cambridge handbook of expertise and expert performance* (pp. 683–704). Cambridge: Cambridge University Press.

Fichman, R. G. (2000). The diffusion and assimilation of information technology innovations. In R. W. Zmud (Ed.), *Framing the domains of IT management: Projecting the future through the past* (pp. 105–128). Cincinnati, OH: Pinnaflex.

Florini, L. (2010). Ethics and information revolution. In L. Florini (Ed.), *Information and computer ethics* (pp. 3–19). Cambridge: Cambridge University Press.

Geertz, C. (1973). *The interpretation of cultures: Selected essays.* New York: Basic Books.

Gero, J. S. (1990). Design prototypes: A knowledge representation schema for design. *AI Magazine, 11*, 26–36.

Giddens, A. (1984). *The constitution of society.* Cambridge: Polity Press.

Giddens, A. (1987). *Social theory and modern sociology.* Stanford, CA: Stanford University Press.

Giddens, A. (2000). *Runaway world: How globalization is reshaping our lives.* Cambridge: Polity Press.

Harpaz, I. (2002). Advantages and disadvantages of telecommuting for the individual, organization and society. *Work Study, 51*, 74–80.

Hulme, M., & Peters, S. (2001). Me, my phone and I: The role of the mobile phone. In *CHI 2001 Workshop: Mobile Communications: Understanding Users, Adoption, and Design, Seattle*, 1–2.

Hulme, M., & Truch, A. (2006). The role of interspace in sustaining identity. *Knowledge, Technology and Policy, 19*, 45–53.

Karwowski, W. (2006). The discipline of ergonomics and human factors. In G. Salvendy (Ed.), *Handbook of human factors and ergonomics* (pp. 3–31). Hoboken, NJ: Wiley.

Kuutti, K. (1996). Activity theory as a potential framework for human–computer interaction research. In B. A. Nardi (Ed.), *Context and consciousness: Activity theory and human–computer interaction* (pp. 17–44). Cambridge, MA: MIT Press.

Latvala, J. M. (2006). *Digitaalisen kommunikaatiosovelluksen kehittäminen kodin ja koulun vuorovaikutuksen edistämiseksi* [Developing digital communication application to advance interaction between home and school]. Jyväskylä: Jyväskylä University Press.

Leikas, J. (2008). *Ikääntyvät, teknologia ja etiikka—näkökulmia ihmisen ja teknologian vuorovaikutustutkimukseen ja—suunnitteluun* [Ageing, technology and ethics—views on research and design of human-technology interaction] (VTT Working Papers No. 110). Espoo: VTT.

Leikas, J. (2009). *Life-based design—A holistic approach to designing human-technology interaction.* Helsinki: Edita Prima Oy.

Leikas, J., & Saariluoma, P. (2008). 'Worth' and mental contents in designing for ageing citizens' form of life. *Gerontechnology, 7,* 305–318.

Leikas, J., Saariluoma, P., Heinilä, J., & Ylikauppila, M. (2013). A methodological model for life-based design. *International Review of Social Sciences and Humanities, 4*(2), 118–136.

Linna, V. (1959–1962). *Täällä pohjan tähden alla* [Under the north star]. Porvoo: WSOY.

Mayr, E. (1998). *The evolutionary synthesis: Perspectives on the unification of biology.* Cambridge, MA: Harvard University Press.

Nardi, B. A. (1996). *Context and consciousness: Activity theory and human-computer interaction.* Cambridge, MA: MIT Press.

Nolan, V. (2003). Whatever happened to synectics? *Creativity and Innovation Management, 12,* 24–27.

Odom, W., Zimmerman, J., Forlizzi, J., Higuera, A., Marchitto, M., & Cañas, J. J., et al. (2013). Fragmentation and transition: Understanding perceptions of virtual possessions among young adults in South Korea, Spain and the United States. In *Proceedings of the CHI 2013: Changing Perspectives, Paris* (pp. 1833–1842).

Pahl, G., Beitz, W., Feldhusen, J., & Grote, K. H. (2007). *Engineering design: A systematic approach.* Berlin: Springer.

Parsons, T. (1968). *The structure of social action.* New York: Free Press.

Saariluoma, P. (1997). *Foundational analysis: Presuppositions in experimental psychology.* London: Routledge.

Saariluoma, P. (1999). Neuroscientific psychology and mental contents. *Life-Long Learning in Europe, 4,* 34–39.

Saariluoma, P., & Leikas, J. (2010). Life-based design—An approach to design for life. *Global Journal of Management and Business Research, 10,* 17–23.

Saariluoma, P., & Oulasvirta, A. (2010). User psychology: Re-assessing the boundaries of a discipline. *Psychology, 1,* 317–328.

Searle, J. R. (2001). *Rationality in action.* Cambridge, MA: MIT Press.

Stahl, B. C. (2006). Emancipation in cross-cultural IS research: The fine line between relativism and dictatorship of intellectual. *Ethics and Information Technology, 8,* 97–108.

Stahl, B. C. (2010). 6. Social issues in computer ethics. In L. Floridi (Ed.), *The Cambridge handbook of information and computer ethics* (pp. 101–115). Cambridge: Cambridge University Press.

Ulrich, K. T., & Eppinger, S. D. (2011). *Product design and development.* New York: McGraw-Hill.

Wittgenstein, L. (1953). *Philosophical investigations.* Oxford: Basil Blackwell.

Young, G. L. (1974). *Human ecology as an interdisciplinary concept: A critical inquiry.* New York: Academic Press.

7

Research and Innovation

> Thinking is the property of the human mind. Scientists engage in constructive thinking: they set up hypotheses and test them, and develop logical chains of arguments to decide which assumptions are facts and which are not. They also create new perspectives for searching out truths by designing more accurate concepts, which allow them to ask new types of questions concerning the states of affairs. However, product design is also a constructive thought process

Product design is also a constructive thought process. Product design can be seen as a form of systematic thinking in which designers generate, evaluate, and specify concepts for technological systems, artefacts, or processes to satisfy the needs of users under specific constraints (Dym and Brown 2012; Dym et al. 2005; Hall et al. 1961; Pahl et al. 2007; Ulrich and Eppinger 2011). Design thinking thus has a creative character (Cross 2004; Eder and Hosnedl 2008; Schön 1983), since design involves finding new solutions or creating new applications for old solutions (Cross 1982, 2001, 2004). Finally, ordinary people, by restructuring their ways of living, turn design insights into innovations (Schumpeter 1939).

Science, design, and innovation are based on the human capacity to create new information contents (Saariluoma et al. 2011). In the exam-

© The Editor(s) (if applicable) and The Author(s) 2016   **207**
P. Saariluoma et al., *Designing for Life*,
DOI 10.1057/978-1-137-53047-9_7

ple of the mining industry, before Savery and Newcomen, water in the mines created problems. However, the construction of primitive steam engines for pumping water out made the problem much more manageable (Derry and Williams 1960). In the beginning of the innovation process, there was little understanding of how to solve the problem with miners' working conditions. Cosimo de Medici failed in his attempts to build a suction pump to raise water from a depth of 50 ft, but with the help of the basic research experiments by Torricelli and Guernicke on the power of atmospheric pressure it became evident that vacuums might have a role in pumps. Finally, the solution of creating a vacuum by steam enabled Savery, Newcomen, and others to develop 'miners' friends', that is, water pumps (Derry and Williams 1960), which dramatically changed the methods of working in mines. Not only designers had changed their way of thinking; the community had also adopted new ideas, and an innovation was born (Schumpeter 1939). This was possible with the help of human thinking and its capacity to create new representations. A web of basic scientific principles made it possible to create ideas, which ended in practical devices for mines. The scientific findings made it possible to construct new design solutions and further develop them into innovations.

Two separate forms of human thinking, design, and research, are important in the development of new HTI products. On the one hand, the designer's thinking *creates* new products, while on the other hand science and human research enable designers to *solve* design problems. As with the natural sciences, human research can be relevant in developing practical solutions. For example, social research on different aspects of ageing in society has an essential role in developing technological solutions to improve the everyday life of ageing people (Sixsmith and Gutman 2013). A critical question is how to organize human research and human life science in order to optimally support HTI processes to improve the quality of human life. In order to develop practical tools for this purpose, it is essential to consider the role of human research in innovative thought processes.

Similarities and Differences

Although the foundations of science and human research,[1] on the one hand and science and design, on the other differ, their approaches have much in common. They have the common goal of improving the quality of human life by means of thinking, although they pursue this goal in different ways and thus often seem to have little to share with each other (Carroll 1997). Human scientists may have difficulty understanding the reasoning of industrial designers, and in the same way many industrial designers have difficulty exploiting the results of human research. One reason for this can be found in their different modes of thinking, that is, their *scientific stance* and *design stance*. In order to understand how human and social research can be effectively linked with design, it is necessary to consider the relationships of these two stances systematically.

The difference between thinking in design and science has been known for a long time in design science (Cross 1982; Cross et al. 1981; Rauterberg 2006). Herbert Simon (1969) in his classic work 'The Sciences of the Artificial', argues that design represents a new kind of science. It is a science of the artificial, and this is typical of schools of medicine, architecture, law, business, and engineering (March and Smith 1995; March and Storey 2008). Since the early 1960s, research into the differences between science and design has been considered in different design contexts (Eder 1998; Iivari 2007; March and Smith 1995; March and Storey 2008; Rauterberg 2006).

Science and design both explicate and describe phenomena, use special terminology to discuss them, like to rely on methods, and construct new ideas and test them (March and Smith 1995; Simon 1969). Thus they have many interaction points. The properties of research and design compared here are their relationships to the nature of theory and practice, the

---

[1] There is an annoying difference in semantic fields and meanings between the English word 'science' and the respective words in other European languages, such as 'Wissenschaft', 'vetenskap', 'ciencia', 'tiede', etc. The latter refer to any form of reason applied in research. For example, such fields of learning as 'literature' and 'history' are forms of Wissenschaft, or vetenskap. However, they are not 'sciences'. When discussing human research, this difference often causes difficulties. Therefore, in this context, 'research' is used as an equivalent term for the European words that describe all forms of investigative activities.

nature of concepts, implicit knowledge as well as the structure of knowledge (Cross 1982; Cross et al. 1981; March and Smith 1995; Hu et al. 2010; Indulska et al. 2009; Simon 1969; Venable 2006).

It is common to think that design is practical and that science is theoretical. However, it is in no way self-evident in which sense science can be seen as theoretical and design as practical thinking. For example, discussion about the movements of the stars long ago demonstrated the importance of the subject for navigation. In fact, this research made it possible to start globalization through colonialism (Bernal 1969). A more modern example would be relativity theory, one of the most abstract theoretical constructions in science, though it has been applied to create power plants and nuclear bombs (Bernal 1969). If Einstein were alive today, he would be fully aware that GPS and many other innovations are based on his theory (Bahder 2003).

The structure of knowledge in science and design differs in many ways. Although concepts and knowledge structures are vital in both research and design, they have different contents. While science works to understand eternally true principles of reality, such as regularities in human life, designers are more concerned with the possible future states of affairs. Questions of science are thus epistemic, while questions of design are praxeological, deontological, and axiological (Anscombe 1957; Boven 2009; Cockton 2006, 2013). Whereas science focuses on 'what is', design focuses more on 'what should be' (Bunge 1959, 1967; Skolimowski 1966; Simon 1969; Saariluoma & Oulasvirta 2010).

It makes sense, however, to also speak about truth and its degrees in design. Design thinking has hypothetical modes. A designer presents ideas, prototypes, models, and concept plans of a more or less hypothetical character as solutions to specific problems (Chandrasekaran 1990; Gero 1990). They are not final solutions, but possible solutions that must be tested or verified against functionality, usability, user experience, and other relevant attributes until it is possible to proceed with the design. The crucial intersection between science and design is thus in the designer's urge to make something possible. To do so, designers must ensure that the design plan is feasible. A precondition of the feasibility of a design solution is that it does not contradict the knowledge of reality in science and human research.

Science helps by asking the right questions and providing realistic answers. This means that it should be possible to manage design and innovation processes by combining design questions with the corresponding research knowledge. In this way, design solutions can have scientific bases and the management of design processes can be organized around asking design questions and providing scientific knowledge to solve them. This kind of design thinking is called explanatory design.

Explanatory Design

A solid ground for assessing what kinds of technological solutions are required for a specific user-interaction design problem can be found by applying human research in interaction design in an explanatory manner. A detailed analysis of user interaction can best be exploited when it is combined with relevant scientific knowledge. It should be possible to explain why one solution is more suitable than another—for example, why a specific solution would be more usable or more robust than other competing interaction solutions.

Explaining is not uncommon in technical thinking. For example, breakage of a car radiator in sub-zero temperatures has a simple technical explanation: in freezing temperatures, the water in the radiator turns to ice, and the force of this expansion is too strong for the radiator to withstand (Hempel 1965). Similarly in medicine, for example, the reasons why different medical methods work are investigated, and this knowledge is used to develop new types of treatments. For example, mechanical forces caused by human bodily movements provide important knowledge about how to treat spinal injuries (Roaf 1960).

In technical design, issues concerning the usage of technologies and the respective design problems can be analysed on the basis of the general laws of nature. Hence, design thinking not only involves explaining something, it also concerns solving problems and predicting different phenomena (Hempel 1965). If ice is the reason why a radiator breaks down, it is necessary to ask how to prevent ice formation. This is a typical question in convergent engineering design. In this case, the problem can be solved by lowering the freezing point by using glycol (Hempel

1965). The explanation is linked to the problem, and solving it directs the designer towards a concrete solution. Explaining, converging, and predicting form a chain of design thought.

There is a fundamental difference between traditional engineering and current HTI design practices. The processes of HTI design are generally dominated by such intuitive procedures as the free and creative generation of ideas, visioning, and user testing. In these, explanatory practices typical to natural science have only a minor role or are even absent, although they could have the power to place HTI design on a more solid foundation (Saariluoma 2005; Saariluoma and Jokinen 2014; Saariluoma and Oulasvirta 2010).

Explanation should be based on science and scientific principles (Fig. 7.1). When designing for life, it is essential to search for explanatory solutions. The main challenge is to unify design problems, scientific truths, and explanatory grounds in a sense-making manner.

In this general explanatory framework, a designer binds the interaction problem and relevant scientific information together and generates a solution on this basis. Each design thought sequence of this type can be seen as a separate explanatory framework, but scientifically grounded design processes are generally characterized by this schema. For example,

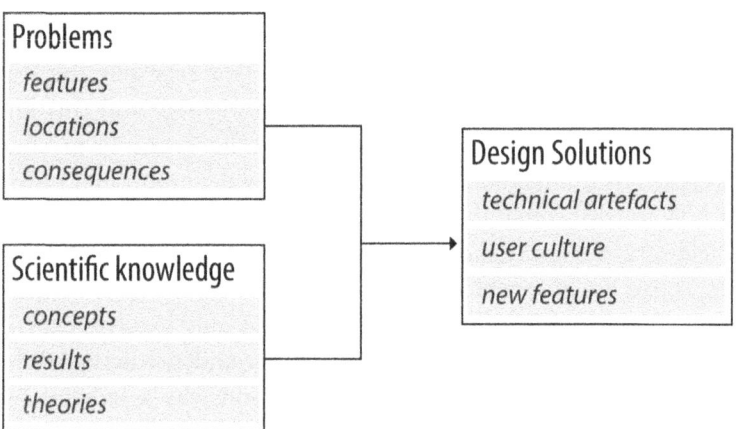

Fig. 7.1 General explanatory framework

theories of visual attention can be used to improve the visibility of icons on a screen. In practice, this means that the device's discriminative character systems must be improved, for example, by organizing colours on the screen. This may require investigating which colours look good and what kinds of icons are experienced as good. The latter problem involves a new explanatory framework that is based on human emotions rather than visual attention.

The core concept of the explanatory process is coherence. This means that the *explanandum* (the phenomenon to be explained) and the *explanans* (the explaining phenomenon) should be coherent. For example, the superiority of GUIs over text-based interfaces can be explained by the pre-eminence of the results of pictorial memory in experimental psychology compared to memorized symbols (Standing et al. 1970). The two phenomena are coherent and consistent. The structure of the explanatory argumentation is similar to Hempel's radiator illustration.

Explanatory frameworks can help designers concentrate their thinking on essential problems in an organized form. When a problem can be transformed into a language that points in a scientific direction, it is possible to enrich it with research knowledge; in this context, coherence plays a significant role. It is essential that the theoretical knowledge and generated solutions are coherent so that there are no contradictions between the design solution and the general base of existing scientific knowledge. For example, it has been known for a long time that discriminate features are crucial in human attention (Gibson 1979; Eimer et al. 1996; van der Heijden 1992, 1996; Schneider et al. 1984). It is evident that discrimination is an important sub-process when people search for information on a display. Explanatory coherence means that the design solutions are consistent with the known principles, results, and theories in basic science. In the case of discriminating between icons, the design solutions have to be coherent with the main truths of visual attention. So, general cognitive knowledge of colour can be used to place the target with a clearly discriminating colour compared to the background to improve the efficiency of a visual search (Treisman and Gelade 1980). The coherence between tradition and the design solution makes it possible to predict whether the solution will work, and to explain why.

Systematization of knowledge, which is the basis of explaining, is also related to falsification (Popper 1959). This means that researchers make predictions based on general principles and laws, against which every scientific truth is continuously tested. If the outcome of the test is positive, the given truths are considered valid. However, if the results do not support the assumptions, the theories must be reformulated (Kuhn 1962).

Testing is also essential in interaction design. However, the outcome of testing can only tell which kinds of solutions will work and which will not; it does not reveal what kinds of ideas would be rational to test. In this sense, tests are essential and can lead to good ideas, but they cannot replace an in-depth understanding of the design problem setting and solution. Frameworks are essential for developing solution ideas and understanding why some solutions work and others do not. They are essential for generating solutions and eliminating suboptimal ideas.

Science and design both have theoretical and practical features. In design thinking today, practicality can only be found by unifying science and design. *Unification* is the key to understanding the function of scientific knowledge in design thinking. It is known that science and design have multiple connections, and that they are at their best when the two stances are unified. However, a precise definition of the connections between the two ways of thinking requires extending the circle of core issues.

Although design thinking is clearly oriented towards solving practical problems in human life, this does not mean that science should be considered impractical. It simply has a different practical orientation: it produces knowledge to be used in practical life. History has illustrated that scientific knowledge has always been superior to intuitive thinking, that is, superior to pseudoscience (Bunge 1967). The less it is necessary to rely on intuitions—and the more detailed design questions that can be asked—the better. However, it is not possible to replace all intuitive elements in HTI with scientific solutions. There is no field of design in which this has been done because design targets new issues, and science does not cover all the practical issues needed in design. Some solutions have to be created with the help of intuition. The following sections examine the extent to which science can replace intuition.

Human thinking begins with problems, which emerge when people have goals but do not know how to reach them (Duncker 1945; Newell and Simon 1972). Defining and setting problems is as important in science as it is in design (Bunge 1967; March and Smith 1995). Yet, the nature of the problems is one of the main differences between science and design. Design problems are normally expressed in terms of performance requirements. The goal of design thinking is to express what the technology being designed is supposed to accomplish, and consequently, to introduce a system with the required performance. In this way, design thinking is consistent with normal human problem solving: it sets goals and searches for intermediate means to reach them (Simon 1969). Scientific problems, in turn, are about *principles* that dominate reality, which can be physical or biological, psychological or social. The principles may concern objects, but also ideas or, for example, social practices or economic activity. Science typically works to unveil the general principles that control reality. Ideally, these principles are thought to be universal and unaltered. Galileo first explained the laws of motion, which are not expected to change over time. Similarly, the principles of the human mind as defined by psychology are anticipated to remain valid in the future. For example, human vision, following its own laws, constructs the experienced three-dimensional world (Rock 1983). Consequently, the principles can be used to solve practical problems independently of time.

Since the main goal of science is to unveil how things are and to describe the prevailing state of affairs (reality), scientific problems have a specific character: they are problems of *knowing*. The solutions to these problems tell us how things are (Bunge 1967). Problems are closely linked with scientific tradition, and they often emerge from testing existent principles.

Design problems are very different from scientific problems. The designer's task is to construct artefacts, information structures, or technologies to help users reach a given goal in their life. For example, when designing an interaction model, the main justification is not that all the solutions are true in the scientific sense, but that the artefact helps people reach their practical goals and satisfy their practical needs (Dieter and Schmidt 2009; Pahl et al. 2007). This is why design problems commonly arise from the recognition of a need that people wish to satisfy.

For example, emotions are an important problem in science and they also have a function in HTI design (Power and Dalgleish 1997; Norman 2004). Psychologists and sociologists work to understand the structure and function of human emotions (Frijda 1986; Power and Dalgleish 1997). They are interested, for example, in the kinds of reactions that emotional words might inspire in people. Designers, for their part, construct emotional features in their systems to make them more appealing and more sellable (Norman 2004).

Thus, there is a logical connection between science and design that is built on the notion of *truth*. Deontological propositions can be considered to have a truth value if there is a possible world in which they are true. Only impossible propositions are false in a deontological sense. This rather obscure claim simply means that designed technologies can work if they can be realized, and they can be realized if the principles upon which they are built are true when the technology is used.

Managing HTI Design Knowledge

In HTI design, innovative thought processes are often more or less unorganized. Because of hectic design cycles, interaction designers do not necessarily have the time to apply systematic methods or use scientific knowledge to construct interfaces (Norman 1986). Although it is common in engineering design to apply the laws of nature and other scientific facts, this approach is unfortunately not often taken in HTI design.

If an organization wishes to exploit scientific knowledge in interaction design processes, it is important to create systematic procedures for doing so, for example, by defining the relationships of relevant design concepts and questions and organizing design processes around them. The concept of usability, for example, opens up a large set of questions and subquestions that can be both general and product specific in nature.

The notion of usability refers to the extent to which a product can be used to achieve users' goals effectively, efficiently and to their satisfaction, and the ways the technology should be designed in order to make it as usable as possible (Karwowski 2006; Nielsen 1993). Thus, to meet this broad design goal and to organize usability and ergonomic design processes in industry, it is also essential to pay attention to such issues as interaction components, controls, and displays for feedback information, contexts of use (Cooper et al. 2007; Galiz 2002; Griggs 1995), and the skills and other properties of users and organizations (Orlikowski 1991, 2001). Using relevant conceptualized questions is especially critical during technology revolutions. For example, without touchscreens, it would not have been possible to formulate design questions concerning this technological paradigm. The way design questions are conceptualized and asked is often of primary importance, but the organization and transparency of these questions can also play a significant role in managing design processes.

Organizing knowledge is vital, as it makes it possible to reuse previous knowledge, as well as guide and manage design thinking effectively. Knowledge management in design has been a topical issue for some time (Gero 1990). A key concept in this discussion is *ontologies*,[2] which are organized sets of domain-specific concepts (Chandrasekaran et al. 1999; Gruber 1993) that describe the most general concepts in a given field; they are widely used in knowledge management. Ontologies can be seen as theories of information contents (Aristotle 1984; Chandrasekaran et al. 1999; Gero 1990; Gruber 1993, 1995) and normally express the information structure of a field (Chandrasekaran et al. 1999). They can, for example, be used to describe an electronic device (Gruber 1993, 1995) and have such content elements as *unit, measure,* and *meter.*

Much of ontology research is focused on formal ontologies (Leppänen 2006; Sowa 2000), which use formal operations and theoretical models. This kind of ontology has no domain-specific contents, but investigates ontologies on a general level. The main idea is that the entities in model sets can be bound to entities in the real world, and in this way the interpretation creates their contents. This book has been concerned with ontologies of HTI; these ontologies are domain- and content-specific ontologies that explicate the main issues of HTI design (Saariluoma and Leikas 2010).

---

[2]It is important to distinguish between general and product-specific HTI design ontologies. General design ontologies describe properties of HTI solutions that are common to all products. Product-specific ontologies concern properties typical to individual products. Only general ontologies are discussed here.

Traditionally, ontologies have mainly been used to describe the structures of some domains as facts. For example, some products have been described as sets of elements and relations. In such instances, ontologies have had the role of information storage and retrieval. However, when considering design as a dynamic thought process, it is more worthwhile to discuss ontology as a *question structure* to be used to manage design problems. Ontologies, in this sense, can be seen as tools for creative thought rather than as information storage. Especially in HTI design, which is influenced by many fundamental concepts and their contents, it is logical to imagine ontologies in the form of questions and answers. They define the information contents of the propositions, which express the domain-specific substance of design problems. Ontologies for HTI design can, thus, be used to generate sets of design questions describing the interaction—that is, the questions that must always be answered when a technology and its relationship to users is designed—and to conduct the HTI design process accordingly.

As concepts, questions and ontologies provide a means of managing corporate thinking and making corporate knowledge explicit. When thought processes are explicit, it is possible to support them, to provide correct knowledge to thinking, to foster innovations, and to move tacit knowledge from one process to another. In this way management can follow the progress of the design process and the content of the design problems that have been solved so far.

From an ontological point of view, HTI design and innovation processes are clustered around four design issues that are intertwined with the major research programs in the field, which were introduced in Chap. 2. These (and their respective design tasks) include:

1. The position of technology in human actions—*design of the role of technology in life*;
2. Functionalities and technical user interface design—*design of functionalities*;
3. Fluent use or ease of use—*design of usability*; and
4. Experience and liking—*design of dynamic interaction*.

These entities are more or less independent of each other and are based on different theoretical discourses and multiple theory languages.

Together, they cover the fundamental ontological questions of HTI design:

1. *What is the technology used for?*
2. *How is it intended to behave and to be controlled?*
3. *How does it fit the users' skills and capabilities?*
4. *How does it produce a motivating emotional experience?*

These principal questions discuss the interaction from the point of view of the human user, and must always be examined when HTI is designed, and must be explicitly or tacitly solved in order to finish the overall design process. The questions are thus task necessary. They have numerous sub-questions characterizing individual products or services. To proceed successfully in the interaction design process, it is vital to consider what the sub-questions are and how they can be explicated. The more accurate and well grounded the ontologies are, the faster and more reliably the design processes can be managed. As the core idea of ontology is to explicate the contents of knowledge, it allows designers to direct routine questions to the appropriate knowledge base. If the interaction information can be easily found this saves time, which leaves more time for creative design issues. Ontologies also enable designers to recall all details of design and benefit from past solutions.

The following section explores these fundamental questions of HTI design and introduces general ontologies for HTI design thinking. The ontologies are not presented sequentially, as it is possible to start the design from any of the main questions. In fact, using the ontologies in an interactive manner makes it easier to proceed with design and innovation processes. Even the parallel use of ontologies is possible and sometimes even recommended, although it may make design management more challenging.

Finally, there is an underlying connection between design ontologies and explanatory design. Design ontologies can be formed around design problems and their respective scientific knowledge bases. Thus good design ontology provides the major task-relevant questions and the respective scientific knowledge bases. Consequently, design ontologies generate explanatory frameworks that connect science and design.

Generic HTI Design Ontology

As already pointed out, the highest-level ontology of HTI design defines the fundamental questions of HTI design processes. The logic is clear: at a minimum, the designers have to be aware of (and solve) the task-necessary questions. Thus, the questions define a set of basic requirements, but at the same time they define a goal for each sub-step in design thinking. The questions cannot define the solutions, but when linked with knowledge of the earlier possible solutions, they can provide the information needed to solve the problems. All design processes are unique, even though the answers could be inherited or transferred from one generation of products or from a certain kind of product to another. In each case, the solutions must be integrated into a unified whole that includes old and new ideas.

At the highest level, HTI design issues are product independent and do not change during the interaction design process. These questions are introduced in Table 7.1 and discussed below.

Table 7.1 The fundamental questions of HTI design ontology and the respective design frameworks

Fundamental design question	Design task	Design framework
What is the technology used for?	The role of technology in life	Life-based design
How is the technology intended to behave and to be controlled?	Control of the behaviour and performance of the technology	Functionality and user interface design
How does the technology fit the users' skills and capabilities?	Being able to use	Usability design
How does the technology produce a motivating emotional experience?	Dynamic interaction	User experience design

What Is the Technology Used for?

What is a technology intended to be used for, in the first place? Why is it used? What is its position in people's life? Answering these questions is a prerequisite for successful interaction design, and calls for an understanding of the life settings that the artefact is intended to support. This is possible with the help of a general notion of 'form of life' that can describe any domain or context of life in relation to technology. Form of life is a system of actions in a specific life setting with its rules and regularities, and facts and values that explain the sense of individual deeds and practices in them.

The LBD process begins with defining a particular form of life and ends with analysing the impact of technology on people's quality of life (fit-for-life analysis). A typical set of ontological questions in LBD is presented in Table 7.2. The questions represent partially iterative and overlapping phases of LBD (see Chap. 6).

The first ontological questions here concern a deep understanding of the structure, functional relations, and contents of the form of life for which the technology is to be designed. The chosen form of life—the *actors and contexts*, their biological, psychological, and socio-cultural

Table 7.2 The ontological questions of 'What for?'

Life-based design	
What is the technology used for?	
Design question	Knowledge required
Definition of a form of life	Human research on characteristics of actors and contexts of a chosen form of life
Definition of the system of regular actions	Analysis of people's goals and related actions based on research and literature on the form of life
Definition of facts and values of life	Analysis of biological, psychological, and socio-cultural preconditions in a given form of life based on research and literature
Innovation of required technologies	Analysis of technology-supported actions and know-how of concept design methodology
Fit-for-life analysis	Empirical analysis of the fit of the planned technology in the task in question, its impact on the form of life; and its capacity to improve the quality of life

preconditions—must be analysed, and its typical *regular actions* must be defined. This analysis can be used to innovate technologies that support people in their actions and practices. People's goals and regular actions give meaning to technological ideas and technology design.

Analysing *facts*—the different determinants of people's lives—makes the everyday contexts that people live in (and the needs that arise from them) understandable. Facts have a set of psychological, biological, and socio-cultural ontological questions of their own. From the point of view of actors they may include, for example, questions concerning gender, education, experience, health, economic status, and work history. From the point of view of action they may include such questions as relevance, structure, processes, and instruments. From the point of view of context they may concern questions of physical, social, psychological as well as informational contexts.

In addition to facts, different *values* that people follow in their form of life are meaningful and necessary for understanding the question 'what for?' One of the main reasons for not adopting and using technology might be that it does not reflect the socio-cultural determinants (values, lifestyles, and mentalities) of the target group. Therefore, values can reveal how people wish to live with technology, not only how the technology can be used. This requires understanding the cultures in which people live.

The next ontological sub-question of 'what for?' concerns *innovation* and the role of new technology in achieving action goals—that is, how technologies can help people. The question is analysed using knowledge of the topical form of life. Here, meaningful rule-following actions are transformed into TSAs. This ontological sub-question provides the logic of how future technology will be part of users' everyday lives, and thus helps give new form to the old form of life.

Logically, new technological ideas must be tested against their suitability for the chosen task. This activity is examined within the question of the *fit for life* of the technology. Fit-for-life analysis studies whether the new artefact or service can really be used in the given life setting, and what kind of impact it would have on people's quality of life. This approach helps ensure that the design outcome has a genuine role in the life of the target user group. The analysis should study both negative and positive impacts—including the practicality, sustainability, and ethics of the design solutions—to reveal the consequences of adopting the new technology. The outcome of this analysis can be rejection of the technological idea or solution, an improvement on it, or acceptance of the technology as part of the form of life of the target user group.

How Is the Technology Intended to Behave and to Be Controlled?

The first task in any technical design process is to define what the technology is intended to do, that is, the behaviour and performance of the artefact. Important questions include how the artefact will affect its environment when it operates, how it is going to change the technical environment, and how the user can control these issues. This helps define the functionalities and technical user interface of the artefact.

As explained above, technical artefacts can reach their goals if they perform correctly (Gero 1990). The performance of the artefact has to be designed so that it can move from an initial state to its expected goal state and have the intended effect. Steering the artefact to the goal state presupposes a number of choices, which can only be made by the user. Therefore, the user has to interact with the artefact. He or she has to select between choices and launch the respective events.

The interactions processed consist of dialogues between the technical artefact—a machine, device, or program—and the human user. Dialogues are uniquely defined as a limited set of signs. Each sign is associated with an event; that is, a process leading to a definite goal state. The human role is to decide between different goal states and the commands that steer the artefact to the desired state. Sometimes dialogues are simple: turning a power switch on activates the electric current in a machine and makes it possible to use it. Dialogues can also be multidimensional, such as when controlling an aircraft. Often these dialogues can be analysed under the concepts of data stream management.

To make the dialogues possible, proper interaction elements in a user interface are required. These can be input elements launching event flows or output elements indicating the states of different processes. In design language they are called, for example, controls, meters, widgets, icons, or

Table 7.3 The ontological questions of 'how to control'

Functionalities and technical user interface design	
How is the technology intended to behave?	
User interface elements	Relevant research and knowledge
Expected goal states and respective functionalities	Task analysis
Event flows: goals and sub-goals	Programming and engineering
Dialogues	Data stream management
Interaction elements, widgets, and controls	Engineering semiotics
Interface architecture	Interface design

text boxes. Their function is to let users issue commands to the artefacts and to inform the users about the state of the artefact. Thus interaction elements are signs in dialogues between machines and people (de Souza 2005) (Table 7.3).

How Does the Technology Fit Users' Skills and Capabilities?

The next fundamental question in HTI design ontology concerns the fit of the technology with users' ability to use it. This problem concerns the human dimension of the user interface and opens up a new set of questions and sub-questions that can be answered with the help of human research and underlying psychological concepts.

In order to guarantee smooth and easy interaction, user interface architectures should explicitly organize dialogues. For example, elements with similar functions should be associated in a sense-making manner. The foundations of understandable user interfaces can be searched from human research—that is, answers to such questions as why a particular architecture is favoured over another. Accordingly, Table 7.4 divides the fundamental question of 'being able to use' into a set of sub-questions regarding smooth and easy interaction.

The overall schema of the sub-questions includes different types of user tasks, described from the human research point of view, and the relevant

Table 7.4 The ontological questions of 'being able to use'

Usability and human user interface design	
How does the user interface fit users' skills and capabilities?	
User behaviour	Relevant human research and psychological knowledge
Perceptual information processing	Psychology of attention and perception
Human motor functioning	Human functioning and psychology of motor movement
Keeping tasks in mind	Working memory and long-term working memory
Learning and remembering	Skills acquisition and long-term memory
Understanding messages	Communication and comprehension, engineering semiotics, psychology of language
Choosing between tasks	Psychology of decision making
Complex and unpredicted processing of tasks	Situation awareness and complex problem solving

psychological and human research knowledge required to discuss these design problems.

Perceptual information processing refers to how people perceive information as relevant when interacting with user interfaces. The main sections of human perceptual information processing, the psychology of perception and attention, are central in solving the interaction problems of making it easy for people to perceive relevant information. Discriminating the target information, such as a warning light or an alarm button, presupposes that the information is discriminable and distinguishable enough, which involves the psychology of perception and attention.

Motor movements have a significant, yet often underestimated, role in usability design. Most user interfaces require the usage of hands, feet, voice, or other motor systems to control the artefact. Users have to be able to set goals for the technology, which is mostly done by using the human capacity to move. Understanding the principles of human functioning and complex motor performances is a prerequisite for good usability design and for developing technology that is accessible to (and usable by) as many people and types of users as reasonably possible.

Memory creates another complex set of user behaviour problems. Memory processes are central in design. Questions such as how to make information easy to remember and recall, how to support remembering information, and how to help people acquire expertise in a particular domain form the core issues in using memory research in interaction design. It is easy to make mistakes in this complex area. For example, many resources were used to develop WAP—Wireless Applications Protocol, which was impossible to realize as its use surpassed human memory capacity. Luckily, graphical interfaces have made it possible for people to use modern ICT networks in this sense.

Communication between people and technical systems is the next human research issue that must be solved. The core knowledge can be found in semiotics (de Souza 2005), which provides principles that can be used to design semiotic systems or codes for interfaces. The psychology of language and meaning explains why some semiotic solutions are better than others.

The psychology and philosophy of thinking create their own discourse within HTI. People have to decide what they want technical artefacts to do for them. Users have to employ artefacts to solve complex problems, and be able to repair them if they get broken. The systems should support these activities in the optimal manner, which requires an understanding of human thinking.

One pressing issue in the development of information societies is demographic change. Ageing is a physiological and psychological phenomenon that involves changes in cognitive capacities such as the functionality of human memory, perception, dexterity of movements, and performance of senses. These factors especially influence the usage of ICT, and create demands for the design of displays, controls, and keyboards. It is necessary to consider these psychological and physiological aspects of life during the design process.

How Does the Technology Produce a Motivating Emotional Experience?

Psychological knowledge of human emotions and motives offers a good starting point for examining the question of emotional experience, which comprises several sub-questions (see Table 7.5).

Table 7.5 The ontological questions of 'emotional experience'

User experience design	
How does the technology produce a motivating emotional experience?	
Psychological tasks	Relevant human research and psychological knowledge
Emotional appraisal	Emotion and appraisal processes
Motives	What motivates people
Individual differences	Psychology of personality and expertise
Interacting groups	Social and cross-cultural psychology

Emotions are important, as they define the human position towards specific issues (Frijda 1986, 1988). In design, the question is not only about positive emotions but also about asking which emotions are relevant in the particular interaction situation. One should be able to feel angry when the cause is irritating enough and happy when experiencing positive actions. Otherwise, the human emotional system does not operate in a rational manner. In interaction design, it is essential to consider how people experience a situation or event that arouses their emotional responses (Frijda 1988; Lazarus and Lazarus 1994). Designers mostly strive to create a positive mood in their clients when interacting with the product. To understand the emotional dimensions of human experience it is necessary to understand human emotions and motivation, which is closely linked with emotions. People often pursue positive mental states, and are therefore motivated to use artefacts that help them do so. The motives for doing something can be complex and long lasting. The modern psychology of motivation offers a sophisticated framework for analysing the motives for using technologies (Franken 2002). The importance of this sub-discourse is obvious: designers need to know why people use some technical artefacts and ignore others.

Finally, good design also requires an understanding of groups and cross-cultural issues. Cultural practices between different user groups may differ, and people may even belong to different subcultures within them. Further, cultural factors with arbitrary signs and meanings may evolve and change. They can in many cases be related to collective values, and may explain why people do not adopt technologies even when—objectively—they could be valuable for them.

Synthetic Design Thinking and Co-design

So far the discussion has mainly focused on analytical design thinking based on the ontological approach. Ontological questions define the elements of analytical design thinking required to set goals and requirements for design and to analyse existing products in relation to a particular form of life.

Yet *synthetic design thinking* is also needed to be able to proceed in product development. It is based on the idea that products are constructed wholes that include human action systems, and that the development process turns product ideas into constructed and working artefacts by implementing different pieces of knowledge together in a meaningful way. This implementation work builds on the analytical (ontological) thinking described above, and necessitates constructive thinking. In the constructive design phase, the role of HTI design ontologies is to again bring forth the relevant issues of designing for life and thus focus designers' minds on developing technology to improve the quality of life. In this sense, analytical and synthetic design thinking are intertwined into a unified construction process that provides designers with synthetic presentations of their ideas (Gero 1990; Goldschmidt 2003; Rosson and Carroll 2002).

Synthetic design is responsible by nature, in that it includes co-design with end users and other stakeholders, which produces a design outcome that is likely to satisfy the needs of the users in both a functional and qualitative manner. To create a successful design outcome, collaborative design activities with stakeholders are needed across the whole span of the design process. These should be organized in such a manner that stakeholders' knowledge and experience can be fully and effectively exploited in design.

Synthetic design can be divided into different phases in which designers can use several types of *co-design activities* and tools to support their work (Leikas et al. 2013). Design for life begins with attempting to improve human life in a particular situation. Following the ontological questions of 'what for', it thus entails an analysis of a certain form of life: the situation in life, the actors and contexts of that situation, and the values that people follow. This knowledge can be gathered from the literature and, for example, interviews with different stakeholders. To understand how new design elements would improve the quality of the form of life in question, common creativity procedures such as brainstorming or synectics are used to generate the first ideas (Nolan 2003; Long 2014).

Quite often, the source of ideas can also be found in the organization's existing designs, design knowledge, and design traditions. A common example of this kind of ideation is *reverse engineering*, in which designers inspect products (either their own or those of a competitor) by 'cutting' them into pieces to understand their logic and to find ways to improve them. The same approach can be used in designing for life, only the focus of the analysis is on how people use the product and how redesign could help the product better serve the users' purposes. Following this approach, the initial idea for developing a good product does not always have to originate from the analysis of life. It can also be technology driven, as long as the designers have an idea that can be trusted, which is discussed with (and accepted by) stakeholders in co-design sessions. In this sense, when designing products, it is not important *where* the idea came from, but *how* the designer proceeds (and where they end up): the design must include all versatile phases of the LBD process, enabling designers to elaborate their ideas. It would be impossible to launch a successful product on the market without a proper understanding of its implementation in and impact on everyday life, its tasks and functionalities, and the harmonized interplay between the user and the artefact.

The outcome of the form-of-life analysis is brought into concept design. In this phase, a *product or service concept* of a technical artefact or service is created. The main functionalities of the artefact, and the main models for its usage in practice, should be clarified at this stage. Personas and scenarios have been found useful in this phase (Cooper et al. 2007; Rosson and Carroll 2002). Personas are models of potential or thought users that produce descriptions of possible users and their goals and intentions. Scenarios can be seen as concrete and work-driven explications of action contexts (Carroll 1995; Rosson and Carroll 2002). In these two methods, the chosen paradigms generate much of the conceptual structure and functional understanding of the product or service concept.

Functional understanding of how a new product will be used leads to a new level of qualitative consideration that searches for the best possible usability and is focused on concrete solutions of input and output devices. Here, typical questions include where the input devices should be placed, what they should be like and how they should be located in relation to each other. Similar considerations with output devices are also necessary.

Questions of usability come to the fore when the functionalities and constructive solutions of the design target have been defined. Following the ontological questions of 'being able to use', usability issues concern such attributes of the product as how important input elements can be discriminated, how they can be recalled, what kinds of interaction levels make sense and how people can be offered the information they need to make sense-making navigation decisions.

When a design concept has been developed, it can be finalized in order to maximize its emotional usability. Concepts such as aesthetics, trust, frustration, and competence are typical problems to be solved at this stage. The goal is to create a product that people trust and like to use. The ontological questions associated with 'like to use' are helpful in solving these problems. They can provide information about what the relevant problems are and how they have been solved in the past. They can also aid in finding effective methods for developing and evaluating design outcomes.

As can be seen, analytical thinking can support synthetic thinking. In synthetic and constructive thinking, ontologies provide designers and developers with tools for thought. They make it possible to inhere such design knowledge and models, which can be stored and used again in other design projects. These models need not be identical to the actual design solutions, but they may suggest ideas that can be used in the search for final solutions.

Sketches, Prototypes, Models, and Simulations

After having the initial ideas about a possible future product, a new synthetic development process begins in which designers create, co-design, test, and revise their plans. Sketches are transformed into low-level prototypes and then into high-level embodied prototypes and, finally, products (Goldschmidt 2003).

With the help of co-design activities and teamwork, development takes an iterative form. Designers may return time and again to the same basic solutions, but every time in more concrete and advanced forms. This kind of 'circulus fructuosis' or *hermeneutic circle* is common in human thinking. It is a natural consequence of constructive activity. Piece by piece, a whole gets its final form.

Sketches, mock-ups, prototypes, models, and simulations are important in communicating ideas with end users and within the design team (Gero 1990; Goldschmidt 2003; Rosson and Carroll 2002). They are also vital in supporting designers' memory in their thinking. The role of sketches and prototypes is to present a synthetic conception. The ontology of design thinking provides tools to analyse all aspects of the idea, to find things that have not yet been considered and provide possible solution models for different issues.

As human information processing capacity is limited, it is good to use different types of sketches and prototypes to support external memory. Prototypes are an excellent means of collecting design ideas, since they allow designers to consider how different interaction elements fit together. Different levels of presentations are used to carry out iterative discussions of the product's functional and qualitative aspects, and enable developers to embody their ideas. This includes defining the forms and role of technology in everyday life, and to the way new technologies can be implemented: what will the particular artefact do, and how will people use it.

Human interactive behaviour is a complex whole, and is therefore difficult to analyse. Laboratory experiments in which only one particular aspect of interaction is analysed are insufficient to cover all relevant problems of interaction design. An additional tool for analysing interaction processes is to conduct a computer simulation to evaluate proposals related any aspect of the design (hypotheses). The simulation is done first by constructing a computer programme that simulates the cognitive mechanisms responsible for behaviours. Then an environmental situation is created that hypothesizes that the user must respond in a certain way. If the simulated user responds as expected, there is evidence to support a design hypothesis. These kinds of simulations are usually performed in the context of either a general cognitive architecture or a cognitive architecture especially developed for simulating interaction processes (Anderson et al. 1997; Kieras and Meyer 1997). Generally, the simulations are made based on an analysis of what is known about user behaviours and the cognitive processes involved in HTI.

Simulation models thus form a specific class of prototypes. They enable designers to consider sequences of actions and events, and to see how the logic of interaction processes operates and how logical it is. Thus computer simulation, despite its costs, often provides a practical tool for interaction designers to see how their assumptions would work, what the problems are, and how they can be solved.

Although the final outcome of a design process may be summarized in a couple of sentences, it may easily require hundreds of small related inventions before it is ready. After having a basic idea of the product or service, and some understanding of everyday life around it, a development process normally proceeds as an iterative process with numerous repetitions. Step by step, the product becomes more specified and closer to the final solution (Rauterberg 2006).

Fit for Life and Ethical Design

Technology design involves the interplay between analytical and synthetic processes. The previous sections provided an overview of how different visualizations, prototypes, models, and simulations can be used in synthetic design processes to collect and test numerous ideas in order to create a working technology. The next question concerns the possibility of fitting the design solutions into people's everyday lives. How is it possible to know that the design solutions are correct and the design ideas valid?

A critical part of the LBD process is fit-for-life analysis. It studies in what forms (and on what terms) technology would be welcomed and adopted in practice. Autonomous and embedded solutions and social media, for example, can only serve new roles in society if they have the strength to bring added value to people's daily lives. As already explained in the context of form of life, understanding social and cultural settings plays an essential role in this context. These refer to values, expectations, and beliefs that influence the way individuals live and how they understand the technology as part of it. Understanding the *current practices* that people follow in everyday life is also essential, as they affect people's engagement with technology on a practical level and how a people and a community can make use of technology.

The holistic nature of LBD sets requirements for the way design hypotheses, solutions, prototypes, and models are tested during design processes. As explained above, a form-of-life analysis focuses on investigating how the new ideas will be accepted and how they improve the quality of life, while *fit-for-life analysis* focuses on the implementation and realization of new design solutions in the context of everyday life. Thus, as a major validation procedure, fit-for-life analysis ensures that the design outcome fits seamlessly into the complexity of the target users' everyday lives. This fundamental phase in LBD brings a new and essential layer to the traditional development discourse.

Standard usability and user experience designs alone cannot provide information on how a particular technological solution can be properly implemented in daily life. They do not focus on analysing how the future product or service will improve practices of life or how it will fit into the complexity of life settings. They are not interested in explaining the role and function of technology in advancing the good life. For this reason, it is essential to use fit-for-life analysis to explore how new technology can be implemented so as to consider all elements of daily life.

Designing a food-delivery service might serve here as an example. This service aims to deliver groceries to people who prefer shopping online. User interface and usability designs naturally ensure that the service has high usability. The user can access the virtual shop and put groceries into a virtual basket. Fluent use of the service when ordering and paying for the purchases is a prerequisite of good service. A positive user experience is guaranteed with the help of an appealing and nicely designed outlook, and a secure service that people can trust and be confident in.

But how to ensure that the service fits seamlessly into the everyday life of the user? Fit-for-life analysis considers more than usability and user experience, and explores how people's everyday lives are constructed around the future service. For example, an online grocery delivery service may work perfectly from a technological point of view, and the user interface may be highly usable. However, if the designers have not considered how the service will fit into the practicalities of the daily life of the client, it will be at risk of failure. For instance:

- If the package is delivered inside a house or apartment:

  - How to ensure that the person is at home at the time of the delivery?
  - What happens if the user is not at home at the time of the delivery?

- If the package is placed at the front door:

  - How to ensure that the neighbour's dog does not find the package?
  - How to ensure that the groceries do not freeze or spoil in extreme temperatures?

Another example could be a service designed for sending greetings to friends. It may work fine for an ordinary day, but the design should also consider people's forms of life and their need to send, for example, Merry Christmas or Happy New Year greetings. If millions of people try to send a greeting at the same time using the same system or application, the service may become overloaded. This should be considered in the form-of-life analysis, and the information should be brought to fit-for-life design.

These two examples reveal only a small part of fit-for-life analysis, but give an idea of the thinking behind this procedural tool, which can be used to test ideas and construction solutions in practice (Leikas et al. 2013). Fit-for-life analysis investigates current practices in everyday life and uses this information to examine the design concept. The aim is to find deficiencies in the design in terms of practicalities of life, and to suggest ways to improve the applicability of the design target. In the analysis, new concepts, sketches, prototypes, and models are critically studied to find the best possible ways to implement the new innovation. The analysis is carried out on both practical and ethical issues of the technological application under design.

Ethical design should be a focal part of fit-for-life analysis. It should cover the identification and assessment of the ethical aspects or implications of the particular design target. Ways of ethical conduct of new product or service types should be examined to ensure success and general acceptance. Here the main question pertains to the interpretation of 'good'. What can be considered good, from whose perspective, and what kinds of choices generate an increase in goodness? In any event, the interpretation of 'good' leads to discussions of quality of life, technology-relevant moral rules, and people's rights and responsibilities. Technologies may have significant consequences for human well-being, and should always be resolved within the design decisions (Bowen 2009). The overall goal should be to advance and defend human values in light of the possible and actual impact of technology (Bynum 2010).

Ethical design means, first of all, conscious reflection on ethical values and choices with respect to design decisions. Second, it entails reflection on the design processes and the choice of design methodologies. Third, ethical design involves what is ethically acceptable. Finally, it must consider what are ethical goals—that is, what constitutes the good of man. All these dimensions must be handled in ways that are ethically valid and acceptable to the users and to community and society in general.

The ethical issues concerning the adoption and use of technology-supported services are raised and solved in a social, political, and economic context. They also arise in the context of use. How the ethical dilemmas are solved depends on the context—that is, the attitudes and views of the different formal and non-formal stakeholder groups involved. Therefore, the ethical issues related to the introduction, adoption, and usage of technology should always be contextualized. The solutions developed have to consider the specific cultural, economic, political, and legal contexts of different societies. People as technology users are heterogeneous in many respects, including their ambitions, education, culture, former profession, family situation, housing, health, and wealth or poverty. Designers must make sure that the design actions will benefit the intended target group without causing detrimental effects on themselves or others.

These issues are profoundly connected to moral values, a matter that has rarely been considered in traditional approaches to HTI design. From this standpoint, the essential research questions include:

- What kind of influence does the technology have on the users' quality of life?
- Does the technology enhance the quality of life of the users better than any other artefact or solution?
- What needs (and whose expectations) should the technology fulfil?
- Who benefits from the technology? Would other stakeholders benefit from it?
- What are the possible alternatives for solving the problem?
- How should the users (direct and indirect) be seen, interpreted, and understood in the design?
- How are the users involved in the design theoretically and empirically?
- Is the main basis for the design answering the users' needs and expectations?
- What are the multiplicative effects of the solution?

It is essential to discuss these questions from the point of view of technology development processes. Hence, ethical issues should be addressed in relation to concept design, test installations, user involvement in design and, finally, fit-for-life analysis. Ethical aspects are also important in innovation design, in implementing the new practice in the society. If ethics are underestimated, there is a risk of distrust and rejection.

References

Anderson, J. R., Matessa, M., & Lebiere, C. (1997). ACT-R: A theory of higher-level cognition and its relation to visual attention. *Human-Computer Interaction, 12*, 439–462.

Anscombe, G. E. M. (1957). *Intention.* Cambridge, MA: Harvard University Press.

Aristotle. (1984). (1.1–15.32) Categories. In J. Barnes (Ed.), *Complete works of Aristotle* (W. Ross, & J. Urmson, Trans.). Princeton, NJ: Princeton University Press.

Bahder, T. B. (2003). Relativity of GPS measurement. *Physical Review D, 68*, 1–18.

Bernal, J. D. (1969). *Science in history.* Harmondsworth: Penguin.

Boven, W. R. (2009). *Engineering ethics*. London: Springer.

Bowen, W. R. (2009). *Engineering ethics: Outline of an aspirational approach*. London: Springer.

Bunge, M. (1959). *Causality and modern science*. New York: Dover.

Bunge, M. (1967). *Scientific research* (Vols. I–II). New York: Springer.

Bynum, T. (2010). The historical roots of information and computer ethics. In L. Floridi (Ed.), *The Cambridge handbook of information and computer ethics* (pp. 20–38). Cambridge: Cambridge University Press.

Carroll, J. M. (1995). Scenario-based design: envisioning work and technology in system development. New York: John Wiley & Sons.

Carroll, J. M. (1997). Human computer interaction: Psychology as science of design. *Annual Review of Psychology, 48*, 61–83.

Chandrasekaran, B. (1990). Design problem-solving—A task-analysis. *Ai Magazine, 11*, 59–71.

Chandrasekaran, B., Josephson, J. R., & Benjamins, V. R. (1999). What are ontologies, and why do we need them? *Intelligent Systems and Their Applications, 14*, 20–26.

Cockton, G. (2006). Designing worth is worth designing. In *Proceedings of the 4th Nordic Conference on Human-Computer Interaction: Changing Roles* (pp. 165–174).

Cockton, G. (2013). Usability evaluation. In C. Ghaoui (Ed.), *The Encyclopedia of Human-Computer Interaction* (2nd ed.). Hershey, PA: Idea Group.

Cooper, A., Reimann, R., & Cronin, D. (2007). *About Face 3: The essentials of interaction design*. Indianapolis, IN: Wiley.

Cross, N. (1982). Designerly ways of knowing. *Design Studies, 3*, 221–227.

Cross, N. (2001). Designerly ways of knowing: Design discipline versus design science. *Design Issues, 17*, 49–55.

Cross, N. (2004). Expertise in design: An overview. *Design Studies, 2*, 427–441.

Cross, N., Naughton, J., & Walker, D. (1981). Design method and scientific method. *Design Studies, 2*, 195–201.

de Souza, C. S. (2005). *The semiotic engineering of human-computer interaction*. Cambridge, MA: MIT Press.

Derry, T. K., & Williams, T. I. (1960). *A short history of technology*. New York: Dover.

Dieter, G. E., & Schmidt, L. C. (2009). *Engineering design*. Boston, MA: McGraw-Hill.

Duncker, K. (1945). On problem-solving. *Psychological Monographs, 58*, 1–113.

Dym, C. L., Agogino, A. M., Eris, O., Frey, D. D., & Leifer, L. J. (2005). Engineering design thinking, teaching, and learning. *Journal of Engineering Education, 94*, 103–120.

Dym, C. L., & Brown, D. C. (2012). *Engineering design: Representation and reasoning.* New York: Cambridge University Press.

Eder, W. (1998). Design modelling—A design science approach (and why does industry not use it?). *Journal of Engineering Design, 9*, 355–371.

Eder, W., & Hosnedl, S. (2008). *Design engineering. A manual for enhanced creativity.* Boca Raton, FL: CRC Press.

Eimer, M., Nattkemper, D., Schröger, E., & Printz, W. (1996). Involuntary attention. In O. Neuman & A. F. Sanders (Eds.), *Handbook of perception and action 3. Attention* (pp. 155–184). London: Academic Press.

Franken, R. (2002). *Human motivation.* Belmont, CA: Wadsworth.

Frijda, N. H. (1986). *The emotions.* Cambridge: Cambridge University Press.

Frijda, N. H. (1988). The laws of emotion. *American Psychologist, 43*, 349–358.

Galiz, W. O. (2002). *The essential guide to user interface design.* New York: Wiley.

Gero, J. S. (1990). Design prototypes: A knowledge representation schema for design. *AI Magazine, 11*, 26–36.

Gibson, J. J. (1979). *The ecological approach to visual perception.* Boston, MA: Houghton Mifflin.

Goldschmidt, G. (2003). The backtalk of self-generated sketches. *Design Issues, 19*, 72–88.

Griggs, L. (1995). *The windows interface guidelines for software design.* Redmond, WA: Microsoft Press.

Gruber, T. R. (1993). A translation approach to portable ontology specifications. *Knowledge Acquisition, 5*, 199–220.

Gruber, T. R. (1995). Toward principles for the design of ontologies used for knowledge sharing. *International Journal of Human Computer Studies, 43*, 907–928.

Hall, A. S., Holowenko, A. R., & Laughlin, H. G. (1961). *Theory and problems of machine design.* New York: McGraw-Hill.

Hempel, C. (1965). *Aspects of scientific explanation.* New York: Free Press.

Hu, J., Chen, W., Bartneck, C., & Rauterberg, M. (2010). Transferring design knowledge: Challenges and opportunities. In X. Zhang, S. Zhong, Z. Pan, K. Wong, & R. Yun (Eds.), *Entertainment for education, digital techniques and systems* (pp. 165–172). Berlin: Springer.

Iivari, J. (2007). A paradigmatic analysis of information systems as a design science. *Scandinavian Journal of Information Systems, 19*(5), 39–64.

Indulska, M., Recker, J., Rosemann, M., & Green, P. (2009). Business process modeling: Current issues and future challenges. In *The 21st International Conference on Advanced Information Systems Engineering, 8–12 June, Amsterdam* (pp. 501–514). Berlin: Springer.

Karwowski, W. (2006). The discipline of ergonomics and human factors. In G. Salvendy (Ed.), *Handbook of human factors and ergonomics* (pp. 3–31). Hoboken, NJ: Wiley.

Kieras, D. E., & Meyer, D. E. (1997). An overview of the EPIC architecture for cognition and performance with application to human-computer interaction. *Human-Computer Interaction, 12*, 391–438.

Kuhn, T. (1962). *The structure of scientific revolutions.* Chicago: University of Chicago Press.

Lazarus, R. S., & Lazarus, B. N. (1994). *Passion and reason: Making sense of our emotions.* Oxford: Oxford University Press.

Leikas, J., Saariluoma, P., Heinilä, J., & Ylikauppila, M. (2013). A methodological model for life-based design. *International Review of Social Sciences and Humanities, 4*(2), 118–136.

Leppänen, M. (2006). *An ontological framework and a methodological skeleton for method engineering.* Jyväskylä: Jyväskylä University Press.

Long, H. (2014). An empirical review of research methodologies and methods in creativity studies (2003–2012). *Creativity Research Journal, 26*, 427–438.

March, S. T., & Smith, G. F. (1995). Design and natural science research on informational technology. *Decision Support Systems, 15*, 251–266.

March, S. T., & Storey, V. C. (2008). Design science in information systems discipline: An introduction to the special issue on design science research. *MIS Quarterly, 32*, 725–730.

Newell, A., & Simon, H. A. (1972). *Human problem solving.* Engelwood Cliffs, NJ: Prentice-Hall.

Nielsen, J. (1993). *Usability engineering.* San Diego, CA: Academic Press.

Nolan, V. (2003). Whatever happened to synectics? *Creativity and Innovation Management, 12*, 24–27.

Norman, D. (1986). Cognitive engineering. In D. Norman & S. Draper (Eds.), *User-centered system design: New perspectives on human-computer interaction* (pp. 31–60). Hillsdale, NJ: Erlbaum.

Norman, D. (2004). *Emotional design: Why we love (or hate) everyday things.* New York: Basic Books.

Orlikowski, W. J. (1991). Duality of technology: Rethinking the concept of technology in organizations. *Organization Science, 3*, 398–427.

Orlikowski, W. J. (2000). Using technology and constituting structures: A practical lens for studying technology in organizations. *Organization Science, 11*,404–428.

Pahl, G., Beitz, W., Feldhusen, J., & Grote, K. H. (2007). *Engineering design: A systematic approach.* Berlin: Springer.

Popper, K. R. (1959). *The logic of scientific discovery.* London: Hutchinson.

Power, M., & Dalgleish, T. (1997). *Cognition and emotion: From order to disorder.* Hove: Psychology Press.

Rauterberg, M. (2006). HCI as an engineering discipline: To be or not to be? *African Journal of Information and Communication Technology, 2*, 163–183.

Roaf, R. (1960). A study of the mechanics of spinal injuries. *Journal of Bone and Joint Surgery, British Volume, 42*, 810–823.

Rock, I. (1983). *The logic of perception.* Cambridge, MA: MIT Press.

Rosson, B., & Carroll, J. (2002). *Usability engineering: Scenario-based development of human-computer interaction.* San Francisco, CA: Morgan Kaufmann.

Saariluoma, P. (2005). Explanatory frameworks for interaction design. In A. Pirhonen, H. Isomäki, C. Roast, & P. Saariluoma (Eds.), *Future interaction design* (pp. 67–83). London: Springer.

Saariluoma, P., Hautamäki, A., Väyrynen, S., Pärttö, M., & Kannisto, E. (2011). Microinnovations among the paradigms of innovation research. *Global Journal of Computer Science and Technology, 11*, 12–23.

Saariluoma, P., & Jokinen, J. P. (2014). Emotional dimensions of user experience: A user psychological analysis. *International Journal of Human-Computer Interaction, 30*, 303–320.

Saariluoma, P., & Leikas, J. (2010). Life-based design—An approach to design for life. *Global Journal of Management and Business Research, 10*, 17–23.

Saariluoma, P., & Oulasvirta, A. (2010). User psychology: Re-assessing the boundaries of a discipline. *Psychology, 1*, 317–328.

Schneider, W., Dumais, S., & Shiffrin, R. (1984). Automatic and controlled processing and attention. In R. Parasuraman & D. Davies (Eds.), *Varieties of attention.* Orlando, FL: Academic Press.

Schön, D. A. (1983). *The reflective practitioner: How professionals think in action.* New York: Basic Books.

Schumpeter, J. (1939). *Business cycles.* New York: McGraw-Hill.

Simon, H. A. (1969). *The sciences of artificial.* Cambridge, MA: MIT Press.

Sixsmith, A., & Gutman, G. M. (2013). *Technologies for active aging*. New York: Springer.

Skolimowski, H. (1966). The structure of thinking in technology. *Technology and Culture, 7*, 371–383.

Sowa, J. F. (2000). Ontology, metadata, and semiotics. In B. Ganter & G. W. Mineau (Eds.), *Conceptual structures: Logical, linguistic, and computational issues* (pp. 55–81). Berlin: Springer.

Standing, L., Conezio, J., & Harber, R. N. (1970). Perception and memory for pictures: Single trial learning of 2560 stimuli. *Psychonomic Science, 19*, 73–74.

Treisman, A., & Gelade, G. (1980). A feature integration theory of attention. *Cognitive Psychology, 12*, 97–136.

Ulrich, K. T., & Eppinger, S. D. (2011). *Product design and development*. New York: McGraw-Hill.

van der Heijden, A. H. C. (1992). *Selective attention in vision*. London: Routledge.

van der Heijden, A. H. C. (1996). Visual attention. In O. Neuman & A. F. Sanders (Eds.), *Handbook of perception and action 3. Attention*. London: Academic Press.

Venable, J. (2006). The role of theory and theorising in design science research. In *Proceedings of the 1st International Conference on Design Science in Information Systems and Technology (DESRIST 2006)* (pp. 1–18).

8

Epilogue: Designing for Life

The main criterion for HTI design is that it should not only concern the development of a technical artefact and the design of the immediate usage situation, but also help illustrate how technologies can advance the quality of human life. People should be motivated to adopt and use technology by the added value it can bring to everyday life to help them accomplish their goals. The question of how much a technology can improve the quality of human life defines the worth of the particular technology.

To be able to improve the quality of the target users' lives, the true value of any technology should be *measured with the concepts of life*. Human life, in all its complexity, should be the starting point for design. Somewhat oversimplified concepts such as 'user needs' cannot adequately examine the true essence of human life actions. In addition to asking what people need, it is also necessary to consider how they can best use technologies, how they can be motivated to use them and for what purposes. For example, as numerous socio-psychological and anthropological approaches have demonstrated, expectations, values, goals, and cultural factors influence people's motives. Technology becomes meaningful through personal and individual symbolic values as well as through social relationships in

© The Editor(s) (if applicable) and The Author(s) 2016
P. Saariluoma et al., *Designing for Life*,
DOI 10.1057/978-1-137-53047-9_8

a person's everyday contexts. People follow the rules of the forms of life that they have adopted (or have been thrown into) to reach their goals and the technologies they use help them reach these goals. Considering different forms of life, technology may enhance, for example, the feeling of belonging to a certain group or to a particular geographical or virtual space. It may also enhance people's feelings of competence, security and self-efficacy, and promote coping in life. Further, it can facilitate people's opportunities to influence decision-making processes through participation and creativity.

As discussed above, LBD thinking—which investigates the HTI design process in concepts pertaining to research into human life—offers the ultimate framework and grounds for designing for life, and thus underlies all HTI design. LBD concepts should belong to the HTI designer's toolbox, from the front-end concept design to evaluating the impact of designed solutions in people's lives. The power of LBD is its holistic nature. Designing for life cannot be based on the natural sciences and mathematics alone. It calls for applying 'human life sciences' (particularly sociology, psychology, and biology of human life, supplemented with other human life sciences) to design processes. Depending on the research target, the corresponding sciences can be ethnography, organizational research and management, philosophy of the mind, education, ergonomics, medicine, neuroscience, or physiology and anatomy. To this list can be added multidisciplinary areas of research such as cognitive science, gerontechnology, occupational therapy, design science, and art design. The key unifying argument is that the problems of HTI design should be conceptualized (and, arguably) supported in concepts and theories developed for analysing human life.

Technology-driven processes in the design of products and services seldom truly manage to support people's goals because they lack the ability to holistically consider people's lives. Therefore the resulting new technologies may be too complex to use (Ramsay and Nielsen 2000), or be anaesthetic, stigmatizing, or somehow ethically problematic. Or they may simply not match the values of the users or fail to bring them any added value. Often, technology-oriented design begins with the creation of new technical artefacts, and only after that is it asked how they could be used (Rosson and Carroll 2002). While there is nothing wrong with this process, it is also possible to turn the design process around and begin with the role of technology in improving the quality of life. Such an approach gives human researchers a better chance of incorporating research-based understanding of human life into designing new technology concepts at an early enough stage, provided that these concepts are given a clear role in technology design processes. This issue has been brought out in such human-driven and human-centred design approaches as goal-directed design (Cooper et al. 2007), contextual design (Beyer and Holtzblatt 1997), and scenario-based design (Rosson and Carroll 2002), which strive to offer a truthful picture of the position of technology in the everyday life of the target population. Such widely accepted and brilliant conceptualizations as personas and scenarios, for example, can be seen as suitable tools to be used in different phases of LBD, because their contents have the power to express the true position of a new technology in life in a realistic and truthful manner. The truth and validity of personas and scenarios depend on how well these descriptions can express different attributes of the form of life of the target population. Superficial and illusory descriptions may lead to problems rather than be of real use.

Many human-centred design approaches include, for example, observing individuals' daily routines in order to understand users' actual needs. This is a good way to start understanding the form of life of the target group. However, merely observing users' needs does not make the design process itself worth conscious (Cockton 2006). As explained earlier, values, for instance, cover much broader and personal contents than the moral values of human welfare and justice, and include non-perceivable values that are based on both personal and cultural concepts. Thus, when designing for the quality of life, it is important to understand what is holistically relevant for people and to try to adapt the design processes to acquire this information early enough and to exploit it effectively in the design. The design may, for example, concern biological changes and the health of ageing people, but it should at the same time focus on cognitive capacity as well as on the values and goals of this target group in order to produce successful design solutions. Therefore, when designing user interfaces and interaction, for example, one should not only focus on the parameters of a screen or input devices from the viewpoint of usability. In holistic design it is also necessary to be sensitive to biological, psychological, and social restrictions as well as the possibilities of the target group. This means considering, for example, such issues as self-efficacy and a sense of people's coherence. In addition, when looking at design challenges on a larger scale from the point of view of LBD, it can be understood that user interfaces are only one piece of the challenge in interaction design. There are also many other factors to consider when striving to use technology to improve the quality of people's lives. These issues arise from people's forms of life and the role of technology within them. For example, log information can be investigated from the point of view of cognitive psychology and the psychology of emotions to see why people stop using some interfaces or why they drop out of e-learning courses. The first question here concerns the cognitive aspects of technology use, while the latter is related to emotional aspects. For example, dropping out of e-learning courses may be associated with different roles, time management, and coping in life in general.

One may argue that all design is life based. When a ship is designed, for example, it is intended to take people from one port to another, and it thus contributes to human life. From this point of view, traditional naval engineering design clearly connects technology with life and can thus be seen as life based. Since all technologies are used in the context of life in some way or another, what can LBD offer that current ICT design practices cannot? There are two important aspects to consider. First, the role of foundational work is to explicate *implicit practices*. The situation can be seen as parallel with linguistics. When working to abstract the structures of a language, something new that has never existed before in that language should be brought out. In linguistics, this paves the way for the construction of grammar—a tool that makes the structure of language explicit. Similarly, explicating important forms of design thinking can improve the understanding of design processes. Although human beings have always had a place in HTI research, it was only after the explication of human-centred design (ISO 1998a, b) that the analysis of human interaction with technologies became more visible. LBD is a similar explicitly defined way of designing HTI. It provides concepts and procedures to help carry out the design process in an organized manner.

The second advantage of LBD over traditional ICT approaches concerns the complex process of HTI design. The reason for explicating intuitions in technology designers' thinking in the area of HTI is *to make tacit practices visible*. It is possible to investigate explicated design procedures systematically, in order to examine their ideal form and ask how they could best be developed. Intuitive and visionary practices, however, cannot be rationally discussed and developed. With the help of new scientific phenomena (and by understanding their behaviour), science and research can provide designers with an opportunity to ground their solutions in facts. For example, experimenting, testing, and inquiry are carefully developed psychological procedures (Cronbach 1984; Kline 1994) that have made it possible to evolve knowledge on usability and hence create efficient methods for usability design (Cooper et al. 2007; Galiz 2002; Griggs 1995; Stanton 2006).

Moreover, abstract principles of science and research per se are not yet practicable. The results and principles of basic research are not significant unless they are applied to a target in life and to the real world. This can be seen in the case of X-rays. Wilhelm Röntgen discovered that it was possible for certain cathode rays to pass through flesh but not bone. Later this knowledge was used in medicine to produce images of bones in tissues (Bernal 1969). Thus the original scientific theoretical work, when applied to solve a practical problem in medicine, led to the creation of an important medical method and to technology for examining bones amid tissue.

Technology can make many things possible. The development and usage of technologies may even generate novel ways of living that contribute to the development of users' ordinary lives. Technologies in different forms of life may—and already have—started to gain new meanings, and people even seem to become emotionally attached to them. In order to make advancements, these emerging walks of (ICT) life should support the form of life that the users in question lead. Therefore, when designing for life, it is essential to ask how a new technological culture changes human life physiologically, mentally, socially, and ethically, and to understand how technological changes could alter the conditions of living for all humankind. If technologies are considered from this holistic and responsible perspective, they have the potential to increase human happiness and the quality of life.

This book has argued that there are four major questions that have to be answered in all HTI design:

1. What is the technology used for?
   - How can the quality of life be improved?
   - What is the role of the technology in life?

2. How is the technology intended to behave and to be controlled?
   - What are the functionalities of the technical artefact?
   - What kinds of controls and feedback instruments should the artefact provide for the users?

3. How does the technology fit users' skills and capabilities?
   - How can the use of an artefact be made possible?
   - How can the technology be made easy to use?

4. How does the technology produce a positive emotional experience?
   - How can people be motivated to use the technology?
   - What will make people like the technology?

The questions are necessary, and they must always be solved, implicitly or explicitly, knowingly or tacitly. It would be impossible to create an artefact without providing it with an appearance and functionalities, and organizing these attributes in some way. It would make even less sense to make a technology without defining a role for it in everyday life.

These fundamental questions are always present in design; they define the parameters of the main design discourses. Seeing the scientific and design process as a system of discourses gives designers the freedom to apply the research ideas that they see as important and helpful in solving their design problems. Understanding the discursive character of the paradigmatic structure also makes it understandable why it is possible to have partially overlapping—and at the same time partially different— ways of examining things. Research and design in technology development is like all human social development activities. The next 'final solution' is to unify different perspectives (Behrent 2013; Foucault 1972; Habermas 1973, 1981, 1990; Sikka 2011).

LBD (designing for life) turns traditional technical design and development thinking upside down—or, to be more precise, places it on its feet. The process of design thinking should focus first on asking how people live, how they wish to live and how the quality of their life could be improved, and only thereafter on what kinds of technologies can serve this goal. Most technological ideas can be explained by 'a human need', but not all technical solutions can be justified in terms of the benefits of the good life. LBD thinking, by defining the contents of relevant knowledge in relation to different forms of life, meets the human needs of a modern society and contributes to improving the quality of life of its citizens. The first question that innovators, designers, and developers following LBD should ask is how they can improve the quality of human life and *design a good life*.

References

Behrent, M. C. (2013). Foucault and technology. *History and Technology, 29*, 54–104.

Bernal, J. D. (1969). *Science in history*. Harmondsworth: Penguin.

Beyer, H., & Holtzblatt, K. (1997). *Contextual design: Defining customer-centered system*. Amsterdam: Elsevier.

Cockton, G. (2006). Designing worth is worth designing. In *Proceedings of the 4th Nordic Conference on Human-Computer Interaction: Changing Roles* (pp. 165–174).

Cooper, A., Reimann, R., & Cronin, D. (2007). *About Face 3: The essentials of interaction design*. Indianapolis, IN: Wiley.

Cronbach, L. J. (1984). *Essentials of psychological testing*. New York: Harper-Collins.

Foucault, M. (1972). *The archaeology of knowledge and the discourse on language*. New York: Pantheon Books.

Galiz, W. O. (2002). *The essential guide to user interface design*. New York: Wiley.

Griggs, L. (1995). *The windows interface guidelines for software design*. Redmond, WA: Microsoft Press.

Kline, P. (1994). *An easy guide to factor analysis*. New York, NY: Routledge.

Habermas, J. (1973). *Erkentniss und interesse* [Knowledge and interests]. Frankfurth am Main: Surkamp.

Habermas, J. (1981). *Theorie des kommunikativen Handelns* [Theory of communicative behavior] (Vols. 1–2). Frankfurt am Main: Suhrkamp.

Habermas, J. (1990). *The philosophical discourse of modernity: Twelve lectures* (Studies in contemporary German social thought). Cambridge, MA: MIT Press.

International Organization for Standardization. (1998a). *ISO 9241-11: Ergonomic Requirements for Office Work with Visual Display Terminals (VDTs): Part 11: Guidance on Usability.*

International Organization for Standardization. (1998b). *ISO-14915: Ergonomic Requirements for Office Work with Visual Display Terminals (VDTs): Part 11: Guidance on Usability.*

Ramsay, M., & Nielsen, J. (2000). WAP usability, Déjà Vu: 1994 all over again. *Report from a Field Study in London.* Nielsen Norman Group.

Rosson, B., & Carroll, J. (2002). *Usability engineering: Scenario-based development of human-computer interaction.* San Francisco, CA: Morgan Kaufmann.

Sikka, T. (2011). Technology, communication, and society: From Heidegger and Habermas to Feenberg. *The Review of Communication, 11*, 93–106.

Stanton, N. A. (2006). Hierarchical task analysis: Developments, applications, and extensions. *Applied Ergonomics, 37*, 55–79.

Author Index

© The Editor(s) (if applicable) and The Author(s) 2016   **251**
P. Saariluoma et al., *Designing for Life*,
DOI 10.1057/978-1-137-53047-9

Subject Index

The manufacturer's authorised representative in the EU is Springer
Nature Customer Service Centre GmbH, Europaplatz 3, 69115 Heidelberg,
Germany. If you have any concerns regarding our products, please
contact ProductSafety@springernature.com

Printed and bound by CPI Group (UK) Ltd, Croydon, CR0 4YY
23/04/2026
02095587-0006